Hypoplastic Left Heart Syndrome

Springer
London
Berlin
Heidelberg
New York
Hong Kong
Milan
Paris
Tokyo

Robert H. Anderson
Marco Pozzi
Suzie Hutchinson

Hypoplastic Left Heart Syndrome

With 95 Figures

Springer

Robert H. Anderson, BSc, MD, FRCPath
Cardiac Unit, Institute of Child Health, University College, London, UK

Marco Pozzi, MD, FECTS
Cardiac Unit, Royal Liverpool Children's Hospital, Liverpool, UK

Suzie Hutchinson, RGN, RSCN
Little Hearts Matter, Birmingham, UK

British Library Cataloguing-in-Publication Data
Anderson, Robert H. (Robert Henry), 1942–
Hypoplastic left heart syndrome
1. Congenital heart disease in children 2. Children–
Diseases–Treatment
I. Title II. Pozzi, Marco III. Hutchinson, Suzie
618.9′212043

ISBN 1852337656

Library of Congress Cataloging-in-Publication Data
Anderson, Robert Henry.
Hypoplastic left heart syndrome / Robert H. Anderson, Marco Pozzi, Suzie Hutchinson.
p. cm.
Includes bibliographical references and index.
ISBN 1-85233-765-6 (alk. paper)
1. Congenital heart disease in children. 2. Children–Diseases–Treatment. I. Pozzi,
Marco. II. Hutchinson, Suzie. III. Title.

RJ426.C64A63 2004
618.92′12043–dc22

2004040150

ISBN 1-85233-765-6 Springer-Verlag London Berlin Heidelberg
Springer-Verlag is a part of Springer Science+Business Media
springeronline.com

Printed in Singapore.

Typeset by EXPO Holdings, Kuala Lumpur, Malaysia
28/3830-543210 Printed on acid-free paper SPIN 10933640

FOREWORD

The introduction of the three-stage Norwood procedure, coupled with neonatal cardiac surgical transplantation emerging as an option for the management of children with hypoplasia of the left heart, has represented one of the great challenges, and I think one of the great successes, in the care of children with congenital cardiac disease over the last 20 years. The surgical and physiological management has not been without its problems. Whilst progress has at times been difficult, it has been steady. Many centres, some represented by the authors in this book, have achieved outstanding results, and have provided inspiration for others to follow. It has, however, taken a long time for the surgical outcomes to be entirely satisfactory, and there is undoubtedly still a long way to go. Whilst early success may now be being achieved, the long-term outlook is still not normal for these children, whose systemic circulation is dependent upon the morphologically right ventricle after the Fontan procedure. Complications of the Fontan circulation are inevitable, and many may need cardiac transplantation in adult life.

Naturally, a newborn and seriously ill baby has a major impact on family life. It is therefore entirely appropriate that the views of the families should have an important place in this textbook. In addition, our own unit, and I know other units, have found the positive support of families with children having hypoplasia of the left heart to be overwhelmingly important to all the health professionals involved in the care of these children.

The authors in this textbook are to be congratulated for bringing together all the important aspects of the care of children with hypoplasia of the left heart, and providing a goldmine of information for their colleagues throughout the world. This is a fine textbook, representing the work of the last 20 years for a group of patients for whom previously there was no hope.

William J. Brawn, FRCS, FRACS
Consultant Paediatric Cardiac Surgeon
Department of Paediatric Cardiac Surgery
Birmingham Children's Hospital
Birmingham, United Kingdom

CONTENTS

Contributors

Robert H. Anderson, BSc, MD, FRCPath
Cardiac Unit
Institute of Child Health
University College London
London
UK

Leonard L. Bailey, MD
Department of Surgery
Loma Linda University
Loma Linda
San Bernardino, California
USA

Akhlaque N. Bhat, FRCS
Paediatric Intensive Care Unit and Cardiac Unit
Royal Liverpool Children's Hospital
Merseyside
UK

Jacqueline C. Blyth, BSc, MSc, ClinPsyD
Clinical Psychology Department
Birmingham Children's Hospital
Birmingham
UK

Edward L. Bove, MD
Section of Cardiac Surgery
Division of Paediatric Cardiovascular Surgery
Department of Surgery
C.S. Mott Children's Hospital
University of Michigan School of Medicine
Ann Arbor, Michigan
USA

John R. Charpie, MD, PhD
Department of Pediatrics
University of Michigan
C.S. Mott Children's Hospital
Ann Arbor, Michigan
USA

Richard E. Chinnock, MD
Department of Pediatrics
Loma Linda University School of Medicine
Loma Linda
San Bernardino, California
USA

Andrew C. Cook, PhD, BSc
Cardiac Unit
Institute of Child Health
University College London
London
UK

Helen Cox, BSc, MBChB, MSc
Human Genetics Division
University of Southampton
Southampton
UK

Dennis C. Crowley, MD
Department of Pediatrics
University of Michigan
C.S. Mott Children's Hospital
Ann Arbor, Michigan
USA

James A. Fitts, MD
Loma Linda University
Loma Linda
San Bernardino, California
USA

Sharon N. Goomany, RGN, RSCN
Paediatric Intensive Care Unit
Royal Liverpool Children's Hospital
Merseyside
UK

Louise E. Hall, BA
Trustee
Little Hearts Matter
Birmingham
UK

Fiona Horrox, BSc, RSCN, RGN
Paediatric Intensive Care Unit
Royal Liverpool Children's Hospital
Merseyside
UK

Suzie Hutchinson, RGN, RSCN
Director
Little Hearts Matter
Birmingham
UK

Joyce K. Johnston, BSN, RN
Cardiac Transplant Department
Loma Linda University Medical Center
Loma Linda
San Bernardino, California
USA

Thomas J. Kulik, MD
Department of Pediatrics
University of Michigan
C.S. Mott Children's Hospital
Ann Arbor, Michigan
USA

Marco Pozzi, MD, FECTS
Cardiac Unit
Royal Liverpool Children's Hospital
Liverpool
UK

Anees J. Razzouk, MD
Division of Cardiothoracic Surgery
Department of Surgery
Loma Linda University
Loma Linda
San Bernardino, California
USA

Gurleen Sharland, MD, FRCP
Fetal Cardiology Unit
Guy's Hospital
London
UK

Audrey Smith, PhD, FIBMS
Honorary Senior Research Fellow
The Cardiac Unit
Institute of Child Health
University College
London
UK

Sue Smith, RGN, RSCN
Paediatric Intensive Care Unit
Royal Liverpool Children's Hospital
Merseyside
UK

Monica A. Stokes, BMBS, FRCA
Department of Anaesthesia
Birmingham Children's Hospital
Birmingham
UK

Aitizaz U. Syed, FRCS
Paediatric Intensive Care Unit and Cardiac Unit
Royal Liverpool Children's Hospital
Merseyside
UK

Janet L. Taylor, RGN, RSCN
Royal Liverpool Children's Hospital
Liverpool
UK

James L. Wilkinson, MBChB, FRCP, FRACP, FACC, FRCPCH
Department of Cardiology
Royal Children's Hospital
Melbourne, Victoria
Australia

David I. Wilson, BA, MBBS, PhD, FRCP
Human Genetics Division
University of Southampton
Southampton
UK

Chapter 1

The Morphology of Hypoplasia of the Left Heart

Audrey Smith, Marco Pozzi, and Robert H. Anderson

The word "hypoplastic" describes the state of an organ that has become stunted in its growth. In this context, hypoplasia of the left ventricle can be the common denominator in a spectrum of cardiac abnormalities, which also involves the left atrium along with the mitral and aortic valves. Subsequent to an early and extensive analysis by Noonan and Nadas,[1] constellations of the lesions to be considered in this chapter have generally become known as the "hypoplastic left heart syndrome." By the standards of today, however, the use of this term is not entirely accurate. This is because, for the geneticists and others, the term "syndrome" is used to define a group of lesions afflicting multiple systems of organs. Within the so-called hypoplastic left heart syndrome, it is only the heart that is involved. We can understand the logic underscoring the stance of the geneticists. Indeed, it is emphasised in Chapter 4 of this volume. For this reason, when describing our own findings in this chapter concerned with morphology, we simply speak of hypoplasia of the left heart. When reviewing the findings of previous authors, nonetheless, we must recognise that most authorities have labelled the entity as a "syndrome," and undoubtedly the majority will continue to do so in the future. In the past, therefore, various combinations of abnormalities with hypoplasia of the left ventricle, often including those with malformations of the aortic arch in the setting of deficient ventricular septation, were identified vaguely as belonging to the so-called hypoplastic left heart syndrome. More recently, it has become customary to include within the syndrome only those hearts with an intact ventricular septum. Yet there are hearts with deficiencies of atrioventricular or ventricular septation that are found with left ventricles as small as those considered appropriately to represent hypoplasia of the left heart. Thus, we will describe the features of some of these hearts with small left ventricles and ventricular septal defects. We will concentrate our attention, nonetheless, on the "classical" variants of hypoplasia of the left heart, in other words, those with an intact ventricular septum.

Most hearts classified in this fashion have either severe stenosis or atresia of the aortic valve. It should then be remembered that, when the aortic valve is stenotic, the left ventricle is not always small. Indeed, very rarely it is also possible to find hearts with aortic atresia when the left ventricle is of normal dimensions, albeit in association with a ventricular septal defect. Hearts with grossly hypoplastic left ventricles can also be found in association with either discordant or univentricular, atrioventricular connections. In these latter combinations, even though the left ventricles may be extremely small, it is not customary to describe them using the term "hypoplasia of the left heart." The term is usually reserved for hearts with the usual atrial arrangement, concordant, or potentially concordant, atrioventricular and ventriculoarterial connections, and when the ventricular septum is intact.

The most severe form, in which the left ventricle is either extremely underdeveloped or cannot be revealed by gross dissection, is the consequence of atresia of both the mitral and aortic valves. Less severe forms may be found when either the mitral or the aortic valve is atretic, and the cavity of the left ventricle is small, but the associated aortic or mitral valve is patent. In this chapter, when reviewing the morphology of the overall group of hearts having hypoplastic left ventricles, we will make reference to previous descriptions, and supplement the earlier findings with our own previously unpublished observations. It is convenient to describe the hearts in separate groups, namely with complete or partial obstruction at the level of either the mitral or the aortic valves, or both.

We should emphasise at this point that, except for those used for the purposes of comparison, the hearts to be described and illustrated will have unequivocal hypoplasia of their left ventricles. This is significant, since many patients are now being encountered in the clinical arena with less severely hypoplastic left ventricles. Some of these patients can be surgically repaired so as to leave biventricular circulations. It remains moot as to whether such patients are appropriately considered as having hypoplasia of the left heart, the more so since it has been suggested that "hypoplastic left heart complex"[2] would be a suitable term for their description. We are not persuaded that such an approach, in morphological terms, is either valid or necessary. From the stance of the anatomist, precise "cut-off points" have yet be established for reliable inclusion of cases in the hypoplastic category. In our opinion, such criterions can only be established using clinical data, and thus far only on a retrospective basis (see Chapter 5). The morphologist can describe the typical anatomy, but cannot provide measurements that will permit living patients to be placed within or without certain categories.

In our account, we will direct emphasis to the morphology of the right ventricle. Most of the variations in the intrinsic pathology of the right side of the heart are inherent across our selected groups. As such, they will be reviewed together. Our description will follow the system of sequential analysis as recommended by Anderson et al.,[3] this being derived from the segmental approaches introduced separately on the one hand by Van Praagh[4] and on the other hand by de la Cruz and Nadal-Ginard.[5] De la Cruz and associates[6] and Eckner and his colleagues[7] also designed tables for comparison of measurements of malformed hearts with the normal arrangements. Elzenga et al.[8] and van der Horst et al.[9] prepared similar tables specifically for analysis of autopsied hearts having obstructive lesions of the aortic arch. We will refer to all these analyses, comparing them again with previously unpublished measurements of suitable hearts from our own archive.

THE EXTERNAL APPEARANCE OF THE HEART AND GREAT ARTERIES

The Heart

Almost all hearts within the spectrum of hypoplasia of the left heart exhibit the usual arrangement of the atrial and thoracoabdominal organs, although isomerism of the atrial appendages is found in less than one-twentieth of hearts with aortic atresia. In the setting of isomerism, the associated abnormalities of the thoracoabdominal organs must be expected, along with the invariably abnormal venoatrial connections. Hypoplasia of the left heart has also been described in association with mirror-imaged atrial arrangement,[10] albeit that we have not encountered such cases. With the usual arrangement, the cardiac mass is left-sided, with the apex pointing to the left. The hearts often appear to be enlarged, but it is the right-sided atrial and ventricular chambers that are most conspicuous. The left ventricle, and often the left atrium, is very much smaller, although sometimes the left atrial appendage can be dilated and hypertrophied. The right ventricle occupies most of the ventricular mass (Fig. 1.1). Some degree of coarctation is to be expected, almost always with hypoplasia of the ascending and transverse portions of the aorta. The arterial duct is always patent in the neonate, and supplies the conduit to the brachiocephalic and coronary arterial circulations, although it can close, with disastrous effect, should the patient survive without palliation.

Key to abbreviations for all figures: AA, ascending aorta; AL, aortic leaflet of the mitral valve; CS, coronary sinus; D, arterial duct; E, endocardial fibroelastosis; IA, innominate (brachiocephalic) artery; ICV, inferior caval vein; IS, isthmus; LA, left atrium; LAA, left atrial appendage; LCC, left common carotid artery; LPA, left pulmonary artery; LPV, left pulmonary vein; LS, left subclavian artery; LV, left ventricle; MV, mitral valve; OF, oval fossa; PT, pulmonary trunk; PV, pulmonary valve; RA, right atrium; RAA, right atrial appendage; RPA, right pulmonary artery; RPV, right pulmonary vein; RV, right ventricle; S, septomarginal trabeculation; SCV, superior caval vein; TV, tricuspid valve.

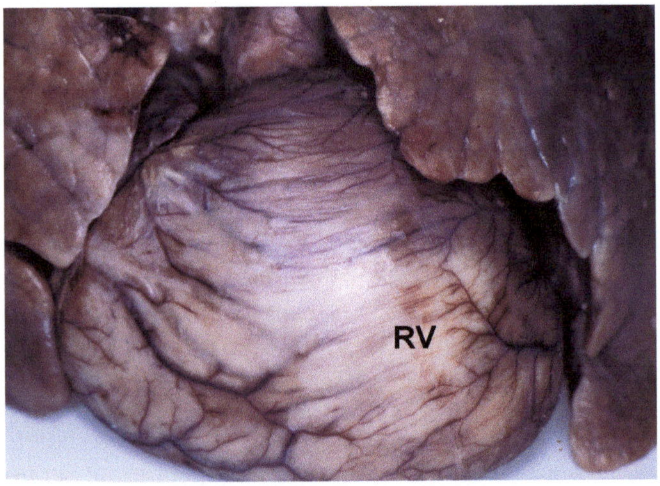

FIGURE 1.1. Hypoplasia of the left heart. The enlarged right ventricle (RV) dominates the ventricular mass.

The Great Arteries

The ascending aorta in hypoplasia of the left heart is small, but variable in size. This is in distinction to those hearts in which coarctation of the aorta is the primary lesion, when the aortic root is more typically close to the normal size, although there is a spectrum extending to true left heart hypoplasia. When the aortic outlet from the left ventricle is atretic, there is a marked tapering of the ascending aortic root as it is traced proximally from the transverse arch toward the base of the heart (Fig. 1.2). At the aortic root itself, the aortic sinuses are frequently prominent, giving rise to the orifices of the coronary arteries. The aortic lumen, therefore, is always patent, even in the setting of aortic atresia, with the coronary arterial circulation supplied retrogradely from the arterial duct. When the aortic valve is patent but critically stenotic, or else formed but imperforate, the aortic

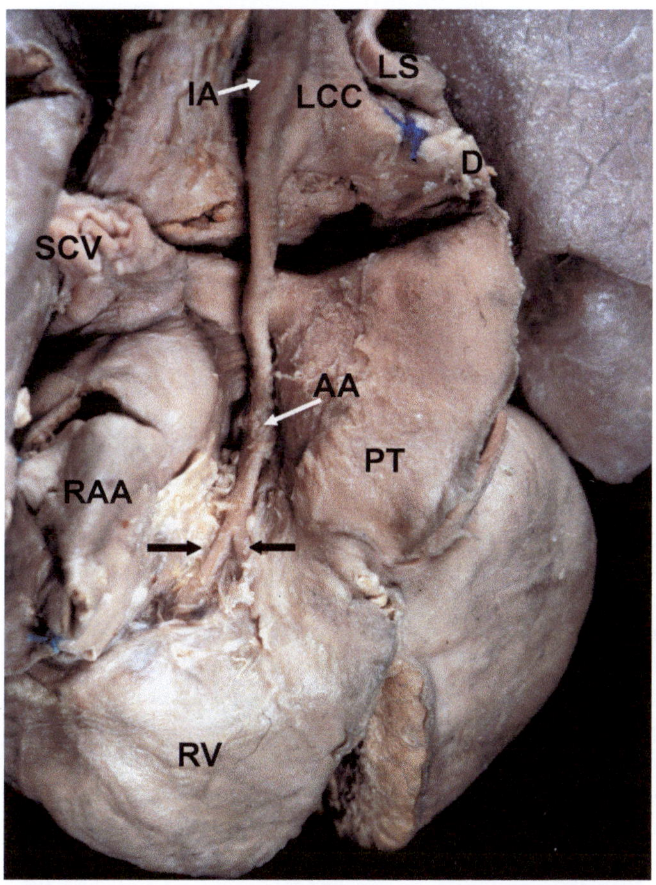

FIGURE 1.2. An oblique view of the thread-like ascending aorta (proximal white arrow) and the origins of the right and infundibular coronary arteries (black arrows) in a heart with aortic valvar atresia and absent left-sided atrioventricular connection. Preductal coarctation of the aorta is present. The isthmus is absent, and the left subclavian artery branches opposite to the arterial duct. The orifice of the left subclavian artery is restricted by the presence of ductal tissue.

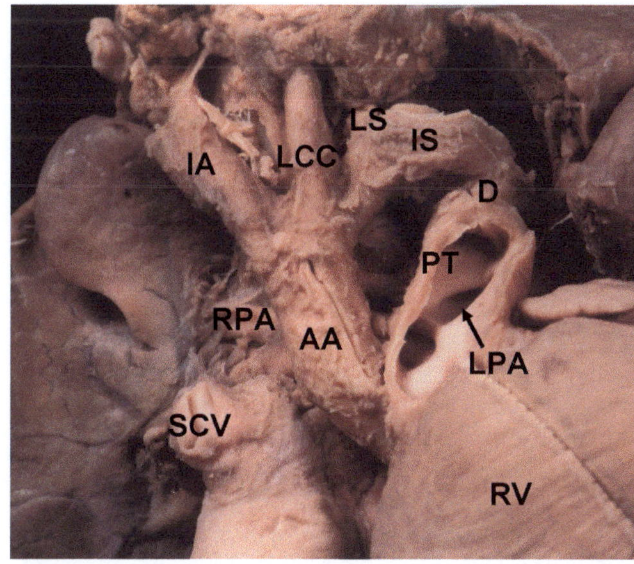

FIGURE 1.3. An oblique view of the ascending aorta in relation to the orifices of the pulmonary arteries and the arterial duct in a heart with small left ventricle but with patency of both the aortic and mitral valves. A long isthmus runs into a preductal coarctation of the aorta.

root is usually proportionately larger (Fig. 1.3), depending on the size of the valve.

The transverse component of the aortic arch in hypoplasia of the left heart is usually wider than the aortic root, although hypoplasia of its segments is a frequent finding. There may then be a tapering of the segments, either toward the isthmus or toward the brachiocephalic arteries. Discrete coarctation, if present, is usually found at the isthmus. Typically produced by an internal shelf, the luminal obstruction may be mirrored by a localised concavity on the external aspect of the aorta, the so-called waist lesion.[11] It has been suggested[8] that the patterns of the aortic arch are predictive for the type of coarctation present. Those that taper toward the isthmus usually present with a preductal coarctation, or in some cases without a coarctation. Those that taper toward the brachiocephalic artery are said typically to have paraductal coarctation. In our experience, paraductal coarctation is the exception rather than the rule, the obstructive lesion in most hearts having hypoplasia of the left ventricle occupying a preductal position.

Our observations also show that the lengths and the circumferences of each of the segments of the aortic root and arch are extremely variable, albeit within similar ranges across all combinations of aortic and mitral valvar atresia and patency (see below). The only differences we noted, which failed to reach statistical significance, showed that the circumferences of the aortic roots tend to be smaller in the setting of aortic atresia when compared with hearts having a patent aortic valve. We will discuss these measurements further in the context of tubular hypoplasia of the aortic arch. Rarely, the transverse arch may give origin to other vessels, such as vertebral or thyroid arteries. Should the left subclavian artery originate opposite the aortic end of the arterial duct, in the setting of discrete coarctation, its origin may be obstructed. The left subclavian artery can also originate distally relative to the arterial duct, although this was not seen in our material. Similarly, the right subclavian artery may be aberrant, taking origin distally to the left subclavian artery, and then passing behind the oesophagus to reach the right side. Again, this pattern was absent from our material.

The pulmonary trunk is either larger than, or within the top end of, the ranges of normal when its circumference is measured in its course from the base of the heart to the level of its bifurcation. Our own measurements were all within similar ranges, irrespective of the specific intracardiac morphology. The right and left pulmonary arteries, however, can take origin from the trunk at different levels from each other, and at varying distances from the pulmonary valve. Their relative positions also vary according to age.

Coarctation of the Aorta

Those describing coarctation have made distinctions between long and hypoplastic segments of the aortic arch as opposed to localised constrictions,[12] and also between atresia of the lumen of a segment of the arch and an imperforate coarctation. We prefer to use the term "coarctation" to describe the localised and discrete deformity of the aortic arch that protrudes into, and partially occludes, the aortic lumen. This may be found as an obstructive shelf, or as a curtain of tissue. The tissue may even encircle the aortic lumen. When seen in association with defects other than with hypoplasia of the left heart, the restrictive coarctation shelf can be found in a preductal, paraductal, or a postductal position. In our series with hypoplasia of the left heart, we found it only in a preductal or a paraductal location, with the preductal examples occurring much more frequently. Detailed descriptions of its exact location tend to be imprecise because of the extreme variability of the extent of the encroachment of the shelf or curtain of tissue into the aorta. The nature of the obstructive lesions has been assessed in hearts without significant associated intracardiac lesions, and in those with hypoplasia of the left heart.[12–15] These studies have shown that preductal coarctation is usually due to the presence of ductal tissue, whereas this is not true of all of those coarctations that are paraductal. Machii and Becker,[16] nonetheless, have described a case with unequivocal hypoplasia of the left heart in which a preductal obstructive lesion did not include ductal tissue, being formed only by medial thickening, with the arterial duct displaying histologic characteristics of persistent patency. Some have found paraductal lesions to be more common in the setting of hypoplasia of the left heart than in hearts with coarctation but without other significant defects,[13] albeit that they were rare in our material. Both preductal and paraductal lesions, nonetheless, have common histological characteristics, suggesting the possibility of a gradual prenatal transition of the lesion from preductal to paraductal location.[8] The portion of the aortic wall proximal to a paraductal obstruction has sometimes been observed to be thinner than the normal, setting the scene for aneurysmal formation. As explained, we have not encountered postductal coarctation in the setting of hypoplasia of the left heart, but von Reuden and colleagues[14] have described three patients with this variant in association with mitral atresia.

Tubular Hypoplasia

When discrete coarctation is associated with hypoplasia of the left heart, it is usually accompanied by tubular hypoplasia of the other segments of the transverse aorta, namely the proximal and distal segments, and the isthmus. The proximal segment lies between the takeoff of the brachiocephalic, or innominate, artery and the left common carotid artery. The distal segment is between the origins of the left common carotid and left subclavian arteries. The isthmus is the segment between the origin of the left subclavian artery and the aortic end of the arterial duct. The narrowing of the lumen of the segments is considered to represent the hypoplasia, but additionally the segments may be lengthened.[8] These abnormalities are usually taken to be the sequel of decreased aortic flow during fetal development. The reduced flow can also influence the structure of the arterial walls. Thus, in the underdeveloped segments of the aorta of patients with tubular hypoplasia, and also in those with aortic atresia, the elastic laminas are densely packed, with a paucity of cellular and supportive connective tissue elements. Nonetheless, the number of elastic laminas in tubular hypoplasia is fewer than in the normal aorta, whereas in patients with aortic valvar atresia, the number of elastic laminas in the hypoplastic ascending aorta is normal. One explanation for this paradox is that there may have been sufficient propulsion of blood into the aortic root to allow normal development before the aortic valve became atretic.[17]

In our autopsied examples of hypoplasia of the left heart, we did not find either a ligamentous segment of the aortic arch, or complete discontinuity between the aortic arch and the descending aorta. The different patterns identified in the morphology of the aortic arch, and in the associated coarctation lesions, almost certainly reflect the pattern of the flow of blood into the aortic root, the arteries supplying the head and neck, and the contribution of the arterial duct. The wide variation of the patterns seen emphasises the importance of the repair of coarctation, and the size of shunts needed for palliation. These features, together with the avoidance of distortion of the pulmonary arteries, are well recognised.[18] The wide variations seen, nonetheless, show that one surgical technique is unlikely to be valid for repair of all forms of aortic coarctation.[11]

Measurements of the dimensions of the ascending aorta, and especially of the transverse arch, are scanty in the literature, although internal diameters have been measured using calibrated probes.[19] Ultrasonographic measurements are usually correlated with age, body weight or length, or body surface area. Alternatively, they are set out as linear correlations, or ratios between different integral segments of the ascending aorta and arch and other major vessels. In this respect, Moulaert and his co-authors[20] defined hypoplasia of the transverse aorta on the basis of decimal fractions. According to them, hypoplasia exists when the diameters of the isthmus, distal, and proximal segments measure respectively up to 0.4, 0.5, and 0.6 of the diameter of the ascending aorta. Elzenga and her co-workers[8] used this method to assess the degree of hypoplasia of the transverse aorta in their series with aortic coarctation and other obstructive lesions of the aortic arch. They also assessed, however, the additional hypoplasia of the ascending aorta and showed that, although many of the segments were hypoplastic, with a single exception all were smaller than the ascending aorta. This method is less useful in quantitating the presence of hypoplasia of the transverse aorta as observed in hypoplasia of the left heart, because the circumferences of the segments of arch are usually bigger than those of the ascending aorta. In an earlier paper, Elzenga and her colleagues[13] had distinguished hypoplasia of the ascending aorta according to its external diameter. They graded the hypoplasia as severe, moderate, or mild according to whether the diameter fell into the ranges of 3 mm or less, 3 to 5 mm, or 5 to 9 mm, respectively. In this work, they did not compare their findings with measurements of the segments of the transverse aorta. Karl and his co-authors[21] assessed the degree of hypoplasia of the aortic arch by matching the diameter of the affected segments of arch against the weight of the patient, in kilograms, plus a factor of one. Their patients had severe hypoplasia or interruption of the aortic arch, but did not have hypoplasia of the left heart. If the diameter measured less than the sum of the other two figures, the arch was said to be hypoplastic.

Our own results are directly comparable with the findings of van der Horst and his colleagues,[9] who chose to make their measurements, as did we, in hearts with hypoplastic left ventricles and aortic atresia. Both sets of data (Table 1.1) confirm the severe degree of hypoplasia of the ascending aorta as compared with normal values. Measurements of the transverse aortic arch in comparison with those of the ascending aorta in

Table 1.1. Hypoplasia of the left heart: includes, with intact ventricular septum, the atretic aortic with atretic mitral valves and atretic aortic with patent mitral valves

	n	*Mean*	*Standard deviation*	*Range*	*Median*
Ascending aorta					
Current work	8	0.7	0.22	0.4–1.0	0.8
van Der Horst	12	0.8	0.3	0.4–1.3	
de la Cruz/normal		1.8			
Eckner/normal		2.1	0.3		
Proximal arch					
Current work	8	1.18	0.11	1.0–1.4	1.2
Distal arch					
Current work	8	1.15	0.15	0.9–1.4	1.2
van Der Horst	12	1.2	0.1	1.2–1.4	
Isthmus					
Current work	8	1.27	0.05	1.2–1.3	1.3

Measurements of circumferences are in centimetres.
Ages: Current work: range 2–21 days, median 5.5 days.
van der Horst: range 1–67 days (including 1 postop), median 5 days.
de la Cruz: range 0–1 month.

hypoplasia of the left heart, however, remain equivocal. van der Horst and his associates[9] did not attempt to assess the degree of hypoplasia of the transverse arch, but their absolute values of the "distal" segment showed that this was generally larger than the ascending aorta. As emphasised, our findings support this conclusion, but the precise degree of hypoplasia remains subjective in the absence of standard values.

The Coronary Arteries

The major epicardial coronary arteries usually occupy the atrioventricular and interventricular grooves. Sometimes these arteries, especially the distal portions, are distended and tortuous, and have been described as having a "corkscrew" configuration.[22] Significantly, the course of the epicardial coronary arteries usually makes it possible to discern the location of the hypoplastic left ventricle, irrespective of its degree of hypoplasia (Fig. 1.4). Lloyd and his colleagues[23] found that measurements of the diameters of the orifices and proximal courses of the coronary arteries showed them to be no different from the normal arrangement. They found little medial fibrosis, although they did record extensive myocardial necrosis. O'Connor and his co-authors,[24] albeit without finding luminal narrowing, have noted thickening of the coronary arteries. In a later study, Baffa and her co-workers[25] estimated the ratios between the thicknesses of the coronary arterial walls and the diameters of the lumens of the right, the anterior and posterior descending, and the circumflex arteries in a group of hearts with hypoplasia of the left ventricle. They revealed that there were no significant increases in these ratios in any of their anatomical subgroups, or in comparisons with their set of control hearts. At the same time, there were no significant differences between the coronary arteries with or without tortuosity. Areas of stenosis or interruption of the coronary arteries were not found. Fistulous communications with the small left ventricle have been described, but are not as frequent or as obvious as seen with hypoplasia of the right heart.[22,24–27] We will discuss these features further in the context of the morphology of the left ventricle with aortic valvar atresia and patent mitral valve.

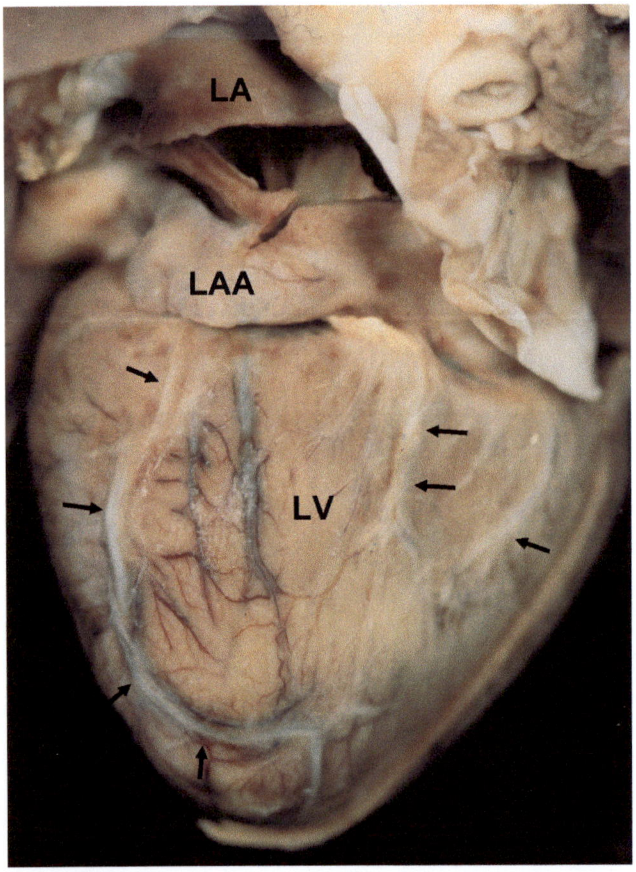

FIGURE 1.4. The left side of the ventricular mass in a heart with atretic aorta and absent left-sided atrioventricular connection. The hypoplastic left ventricle is delineated by the epicardial coronary arteries (arrows).

INTERNAL MORPHOLOGY

The Left Side of the Heart

Following the sequential approach, we can chart the effects of the basic flow of the blood through the left side of the heart, beginning with the left atrium.

The Left Atrium

The left atrium is usually small, but possesses all its components except when there is a totally anomalous venous connection. Often it exhibits whitish thickening of the endocardium, so-called endocardial fibroelastosis. It generally receives all four pulmonary veins, which sometimes may be restrictive. Although anomalous connection of the pulmonary veins has been reported, the more usual anomaly involving the pulmonary veins, seen in perhaps one-tenth of cases, is the finding of an anomalous channel that runs from either the pulmonary venous component of the left atrium, or from one specific pulmonary vein, to the systemic venous system. This potential "overflow" for the restricted left atrium is called the "levoatrial cardinal vein."[28] Another potential overflow, found in association with mitral atresia, is a fenestration, or fenestrations, between the left atrium and the coronary sinus. Partitioning of the left atrium, so-called cor triatriatum, is a rare finding, but occasionally a partial supravalvar mitral ring is present. Viewed from the posterior aspect of the heart, it is possible in some cases to appreciate a leftward deviation of the flap valve of the oval fossa. When found, the anterior edges of the flap are attached closer to the mouth of the left atrial appendage and the connections of the left pulmonary veins to the left atrium (Fig. 1.5) than is seen in the normal situation (Fig. 1.6). Such malalignment of the primary atrial septum can diminish the size of the left atrium. More usually, the valve of the oval fossa is either deficient or probe patent. It may also be thickened or aneurysmal, then producing a restrictive interatrial communication. Otherwise, the flap valve of the oval fossa is often of sufficient dimensions to overlap the infolded anterosuperior rim of the oval fossa. In about one-tenth of cases studied at autopsy, the flap valve is firmly fused to the rim, producing an intact atrial septum. The restrictive or intact atrial septum may have a serious impact on palliative surgery and transplantation, the result being that some neonates may have pulmonary vascular disease at birth.[29] The pulmonary vascular disease is presented histologically as medial hypertophy or peripheral extension of muscle of the pulmonary arteries, leading to pulmonary arterial hypertension, accompanied by arterialisation of the pulmonary veins consistent with venous hypertension,[30] and also severely dilated lymphatic vessels.[31] In an extensive study of infants with hypoplasia of the left heart and intact atrial septum, Rychik and his colleagues[31] evaluated the relationship between the atrial morphology, the pathways for left atrial decompression, and the pulmonary pathology. They identified different types of left atrial morphology in these hearts, which they suggested could be used as a marker for the severity of pulmonary vascular disease. It is interesting that lymphangiectasia has been reported in a baby with hypoplasia of the left heart, intact atrial septum, and deletion of chromosome 18.[32] The effects of the intact or restrictive atrial septum in hypoplasia of the left heart is discussed elsewhere in this volume by Andrew C. Cook (see Chapter 2)

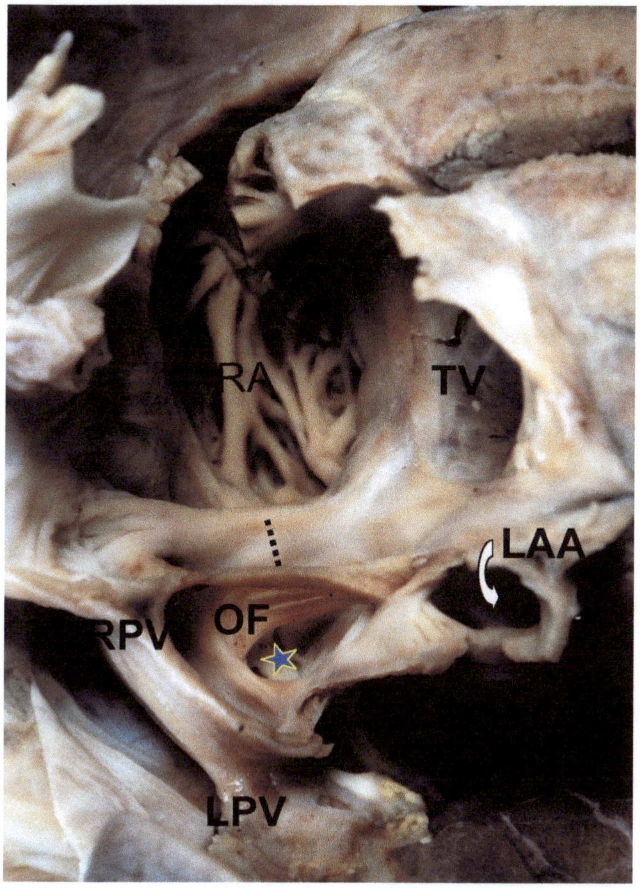

Figure 1.5. The left atrium of a heart with atretic aortic valve, absent left atrioventricular connection and a vestigial left ventricle. The flattened, superior rim of the oval fossa is shown by a dotted line. There is leftward deviation of the flap valve of the oval fossa, which impinges almost upon the opening of the left pulmonary veins and the left atrial appendage. This reduces the size of the left atrium. A defect of the oval fossa is present (asterisk).

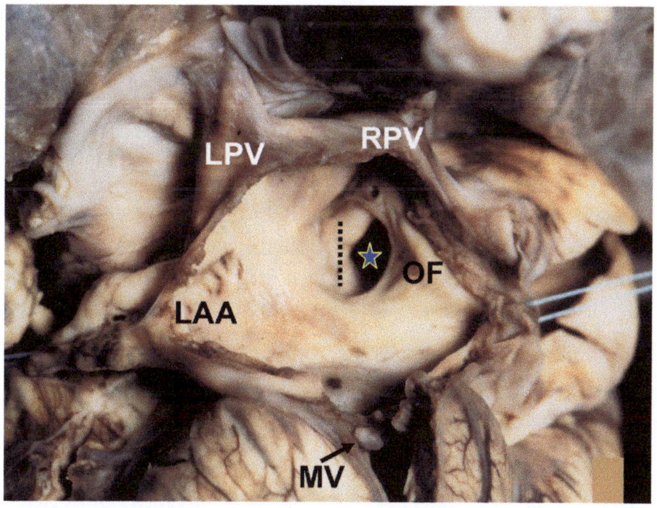

Figure 1.6. The left atrium in a case with hypoplasia of the left heart in which there is no deviation of the atrial septum toward the left. The flap valve of the oval fossa is contiguous with the superior rim of the oval fossa (dotted line), in the usual position, near to the opening of the right pulmonary veins. A large defect of the flap valve is present (asterisk).

The Mitral Valve, Left Ventricle, and Aortic Valve

We have not examined the histopathology of the ventricular walls in our series of cases with hypoplasia of the left heart, but others have found evidence of ischaemia, necrosis, fibrosis and infarction in the left ventricles, without significant differences between the subgroups.[25] Endocardial fibrosis, however, is significantly increased in hearts with patent mitral valve and aortic atresia, this being confirmed by our own macroscopic findings. The main anatomic differences between the various cases properly classified as hypoplasia of the left heart become apparent when examining the nature of the left-sided atrioventricular connection.

Patent Mitral Valve with Aortic Valvar Atresia, Intact Ventricular Septum, and Hypoplastic Left Ventricle

This is a well-defined and discrete combination of abnormalities. It is found as commonly as are the hearts with either absent or imperforate mitral valves, aortic atresia, and intact septum (see below). When the mitral valve is patent, it is miniaturised and usually supported by equally miniaturised, but abnormal, papillary musculature (Fig. 1.7). Often the valve is grossly dysplastic, its tendinous cords being inserted directly into the ventricular endocardium. The valve itself usually has two leaflets, but the intercordal spaces may be reduced in number. Double orifices have been reported.[33] The circumferences of the valves always measure well below the normal range of values. The underlying left ventricle is correspondingly small, and is sometimes described as a "peach-stone ventricle." This is because fibroelastosis of the left ventricle is to be expected when the mitral valve is patent (Fig. 1.7). The lining of the ventricle is pearly white and smooth-walled, reminiscent of the inside of a peach after removal of the stone. This may be indicative of multiple ventriculocoronary arterial connections, as shown by Baffa and her co-authors,[25] who found coronary-cameral communications in almost one-third of their extensive series with this combination of lesions. They also saw these abnormal vessels in about one-twentieth of hearts with both mitral and aortic valvar atresia. On the other hand, in the study of O'Connor and colleagues[24] examining microscopically hearts in which the aortic valve was either atretic or hypoplastic, but the mitral valve was patent, multiple ventriculocoronary arterial connections with associated endocardial fibroelastosis were present in all cases. These also had consistently thick-walled left ventricles. None of these pathological features were present in two hearts examined by O'Connor and colleagues[24] in which there was aortic atresia but in which the mitral valve was atretic. O'Connor and colleagues[24] suggested that the ventriculoarterial patterns represented the persistence of an embryonic situation that allowed for the egress of blood from the obstructed left ventricle at high

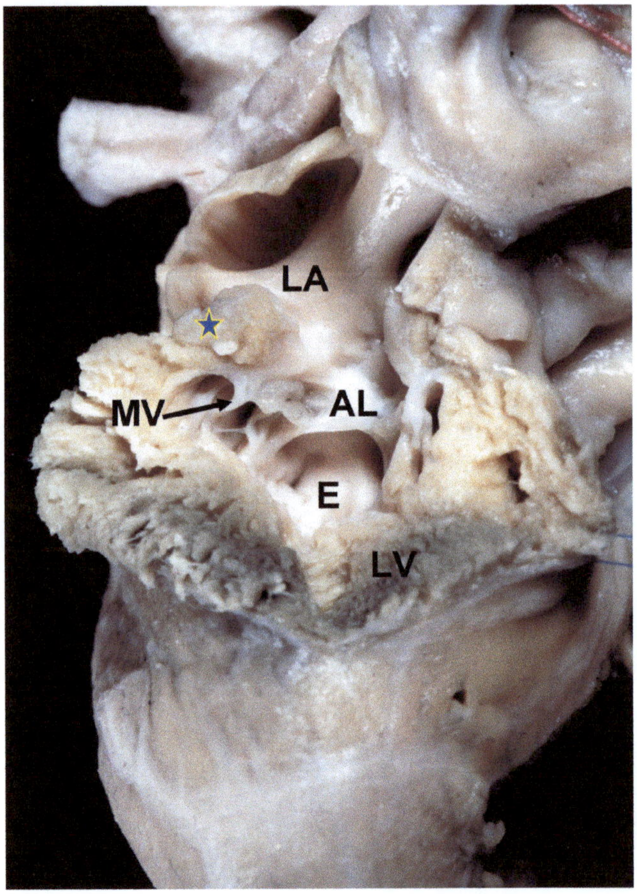

FIGURE 1.7. The mitral valve in this heart has been cut across so that the left ventricle and part of the left atrium may be exposed. The mitral valve is miniaturised and dysplastic and the ventricular cavity is lined with whitish endocardial fibroelastosis. An accretion of accessory valvar tissue is present on the atrial aspect of the mural leaflet of the mitral valve (asterisk). The aortic leaflet of the mitral valve shows an arcuate disposition and the papillary muscles are poorly developed. Some of the cords are attached directly to the wall of the ventricle. Aortic atresia is present. The right ventricle of this heart is shown in Figure 1.19.

pressure. Freedom[22] has commented that such changes could limit surgical attempts to produce a functional left ventricle, but thus far there is no evidence from clinical experience to substantiate this prognostication.

Patent Mitral Valve with Imperforate Aortic Valve, Ventricular Septal Defect, and Variable Hypoplasia of the Left Ventricle

When an imperforate aortic valve is found in combination with a ventricular septal defect, the annulus of the mitral valve and the length of the left ventricle are variable in size, sometimes being small, but at other times falling within the normal range. The mitral valve is usually dysplastic. The imperforate aortic valve can itself vary considerably in size, from being of nearly normal diameter down to a few millimetres in diameter.[34] In these hearts, the distribution of the anterior interventricular coronary artery is likely to be similar to that of the normal heart. An enlarged and dysplastic pulmonary valve has been observed with this combination of anomalies. The ventricular septal defect may be perimembranous (Fig. 1.8) or muscular. We did not include any cases of hypoplasia of the left heart with atrioventricular septal defect in our series.

Patent Mitral Valve, Patent Aortic Valve, and Intact Ventricular Septum, with Variable Hypoplasia of the Left Ventricle

This combination of lesions produces a more heterogeneous picture than does the subset with patent mitral valve, aortic atresia, and intact ventricular septum. The circumferences of the mitral valves, and the lengths of the left ventricles, may either be just within, or just below, the lower range of normal values. The mitral valves themselves, nonetheless, are frequently dysplastic

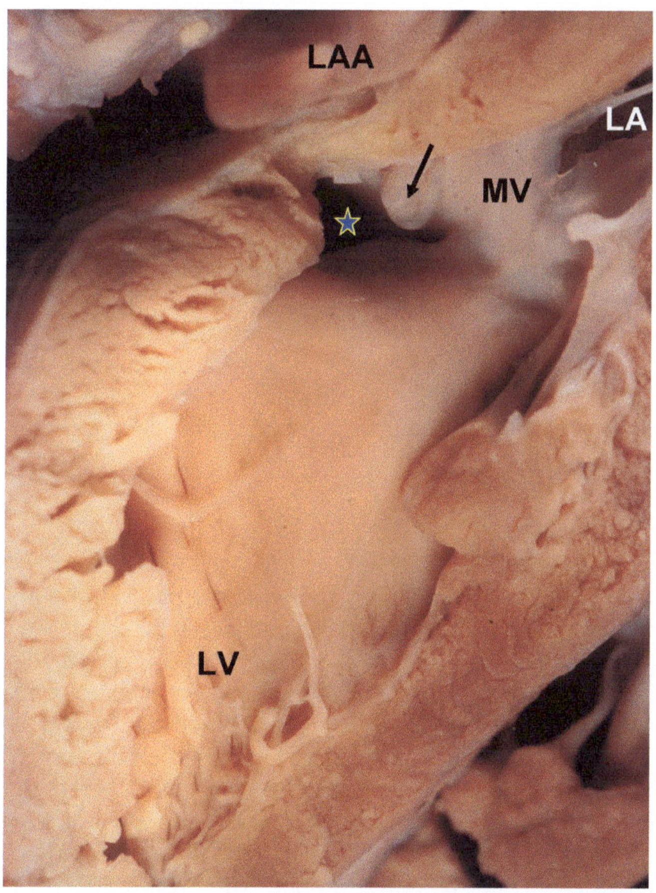

FIGURE 1.8. The left ventricle of a heart with an imperforate aortic valve (arrow), a small, dysplastic mitral valve, and a perimembranous ventricular septal defect (asterisk).

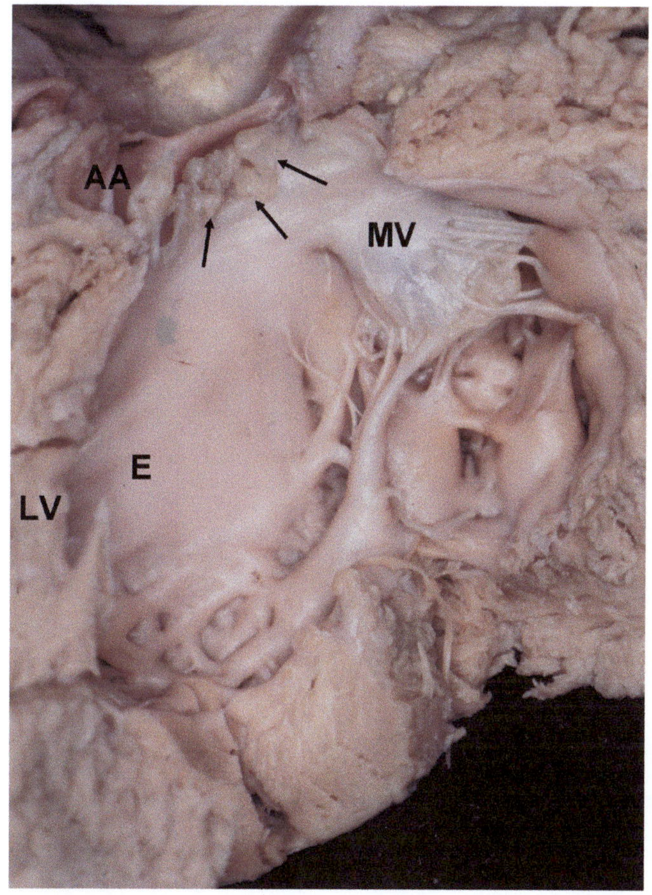

FIGURE 1.9. The thick-walled, left ventricle of a heart with critical aortic valvar stenosis (arrows). The mitral valve is dysplastic and the endocardium is thickened by a layer of fibroelastosis.

and stenotic. The aortic valves may be small, but are often normally formed. This group also includes the lesion known as critical aortic valvar stenosis. In this setting, the aortic and mitral valves are patent, but the left ventricles show extremely variable morphology. Usually they are small, with small dysplastic mitral and aortic valves and endocardial fibroelastosis (Fig. 1.9). In some hearts, nonetheless, the left ventricles fall into the normal ranges, while some are even larger than normal. Thus, there is a spectrum of morphology to be seen with critical aortic valvar stenosis. Those with very small left ventricles are unequivocally examples of hypoplasia of the left heart. Others with patent mitral and aortic valves, intact septum, and larger ventricles fall outside the received perception of this classification (Fig. 1.10). As yet, no consensus has been reached on the criteria for inclusion or exclusion.[2] As we have already emphasised, such criteria will not come from anatomic studies, but only from careful examination of the clinical findings in patients who do or do not survive attempted

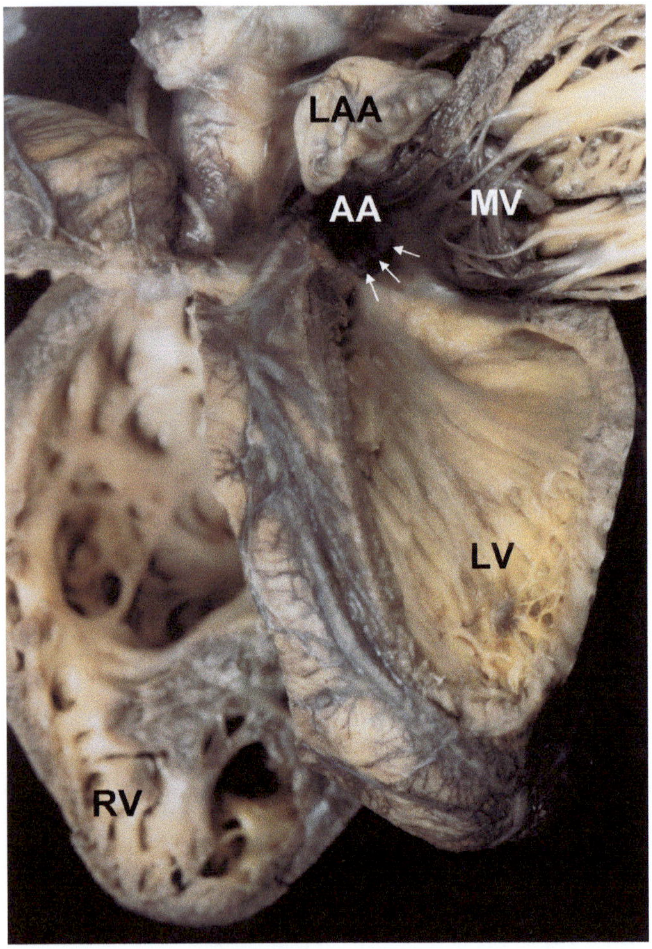

Figure 1.10. The thin-walled left ventricle of a heart with critical aortic valvar stenosis (arrows) in which the length of the ventricle is normal.

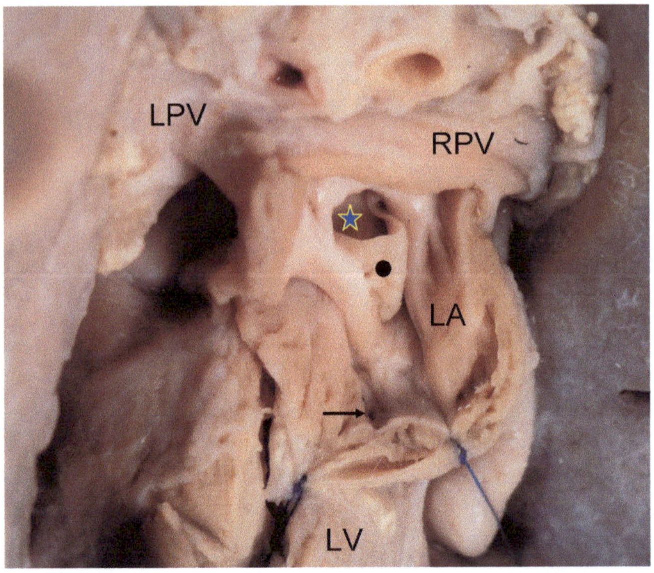

Figure 1.11. The left atrium has been opened in this case to show that the mitral valve is represented by only a small dimple in the muscular floor of the chamber (arrow). Coincidentally the picture shows that the atrial septum is deviated leftward toward the opening of the left pulmonary veins. An atrial septal defect of the oval fossa is present (asterisk) and the flap valve is aneurysmal (black dot).

biventricular repair. As already discussed, this distinction, at present, is made only in retrospect. In our opinion, the fact that some patients can successfully be treated by biventricular repair[2] does not justify the creation of a new category of "hypoplastic left heart complex."

Absent or Imperforate Mitral Valve with Atretic Aortic Valve, Intact Interventricular Septum, and Hypoplastic Left Ventricle

This combination of malformations produces a well-defined group of hearts. The aortic valve is atretic, although it may be formed but imperforate in some instances. The mitral valve is often unidentifiable, but can be represented either by a dimple in the floor of the left atrium (Fig. 1.11), or by a small valvar membrane. In a small proportion, the valve is better formed, with underlying tension apparatus, but still imperforate. In those in which the left atrioventricular connection is absent, the left ventricle may not be identifiable on gross dissection, even though the descending epicardial coronary arteries delineate the potential site of the hypoplastic left ventricular chamber. In this type of anatomy, histological studies may reveal a fibrous connection between the left atrium and the underdeveloped ventricle.[35] In these cases, if performed with care, gross dissection may reveal the vestigial cavity, lined with endocardium, and buried in the left-sided parietal myocardium.

Absent or Imperforate Mitral Valve, with Patent Aortic Valve, and Ventricular Septal Defect

This combination of malformations is often found with concordant ventriculoarterial connections. Alternatively, there may be overriding of the aortic valve across the ventricular septal defect, or the aortic valve may be committed predominately to the right ventricle, giving a double outlet connection. In all of these combinations, the circumference of the aortic valve tends to be small. The left ventricle may be of normal dimensions. In these circumstances, it is common to find fibrous subaortic obstruction as a consequence of proliferation of aneurysmal or accessory tricuspid valvar tissue through the ventricular septal defect. In our material, the combination of patent aortic valve with absent or imperforate mitral valve, ventricular septal defect, and concordant ventriculoarterial connections was found most frequently when the aortic valves, and the left ventricles, measured well below the lower end of the normal ranges. Furthermore, the left ventricles typically had thin, nonhypertrophied walls.

The Right Side of the Heart With Hypoplasia of the Left Heart

The Right Atrium

The right atrium is frequently dilated, and its walls are usually thickened. The superior and inferior caval veins, and the coronary sinus, are connected to the right atrium in the usual way and, rarely, a persistent left superior caval vein may drain to the right atrium via the coronary sinus. The morphology of the atrial septum is extremely variable and has been discussed previously in more detail in relation to the morphology of the left atrium. In approximately one-tenth of cases, the septum is intact. More commonly, the septum has either a deficient or a probe patent oval fossa, but the communication may be restrictive as the result of thickening or aneurysmal dilation of the flap valve (Figs. 1.11 and 1.12). When deficient, there is an obvious interatrial communication across the oval fossa. There can be malalignment between the anterosuperior edge of the flap valve and the corresponding muscular rim of the oval fossa, which is frequently abnormally broad and flattened horizontally. This feature produces an interspace in the right atrium that is directed toward the left atrium (Fig. 1.13). As discussed previously in the description of the left atrium, this is the arrangement that carries the flap valve of the oval fossa closer to the left atrial appendage and to the connections of the left pulmonary veins (Figs. 1.5 and 1.11) than in the normal heart, thus diminishing the size of the left atrium.

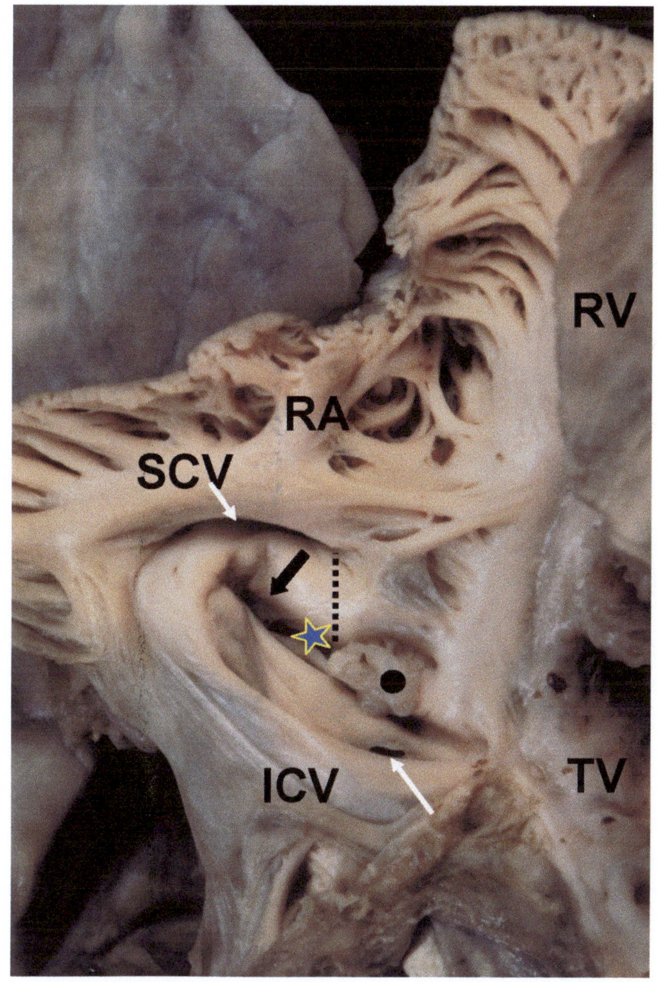

Figure 1.12. The right atrium has been opened to demonstrate an aneurysmal flap valve of the oval fossa from the right side (black dot). This heart has an atretic aortic valve but a patent mitral valve. The superior limbus of the atrial septum is flattened (dotted line) and the flap valve is recessed toward the left (black arrow). A defect of the oval fossa is present within the recess (asterisk). Another small defect is present (long white arrow).

The Tricuspid Valve and the Pulmonary Valve

The annulus of the tricuspid valve in our series usually measured within, or even above, the upper range of normal. Generally the leaflets are normal in appearance, but they may be dysplastic, especially in the presence of a ventricular septal defect. When the ventricular septum is intact, the tricuspid valve has been found to be dysplastic in one-third, myxomatous, thickened, and redundant in one-quarter, and bifoliate in one-eighth of cases.[36,37] Sometimes, because of "remodelling" of the right ventricle, the cordal attachments to the tricuspid valve will have become elongated. Dilation of the right ventricle may be present when the tricuspid valve is supported only by the medial papillary muscle and one other muscle, giving an eccentric "parachute" arrangement. The pulmonary valve is usually larger than, or within the top end of, the range of normal values. Pulmonary valves with two leaflets, or with prolapsing leaflets, have been reported.[38] In a small minority of hearts, the valvar leaflets can be thickened and rolled, as seen in one case from our material.

The Right Ventricle: The Ventricular Septum, the Septomarginal Trabeculation, and Other Muscular Trabeculations

The inlet and outlet dimensions of the cavity of the right ventricle are either within or above the range of normal.

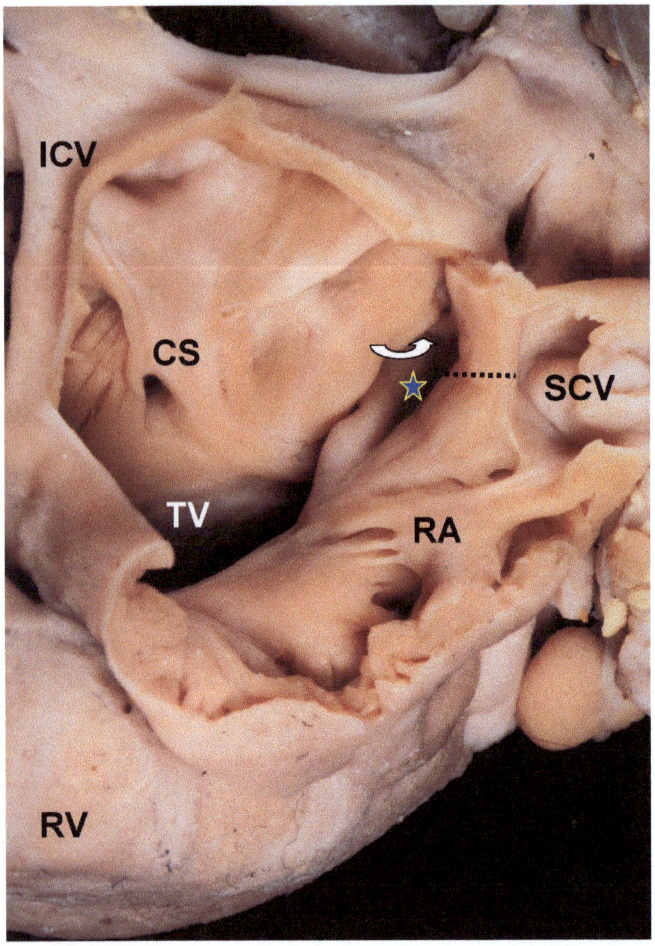

FIGURE 1.13. This view inside the right atrium demonstrates more clearly, when compared with Figure 1.12, that the superior rim of the oval fossa (dotted line) is flattened and that the flap valve is deviated toward the left (curved arrow). An interspace (asterisk) between the two structures is more obvious. In this heart, the aortic valve is atretic and the mitral valve is represented by only a dimple. A left ventricular cavity could not be demonstrated.

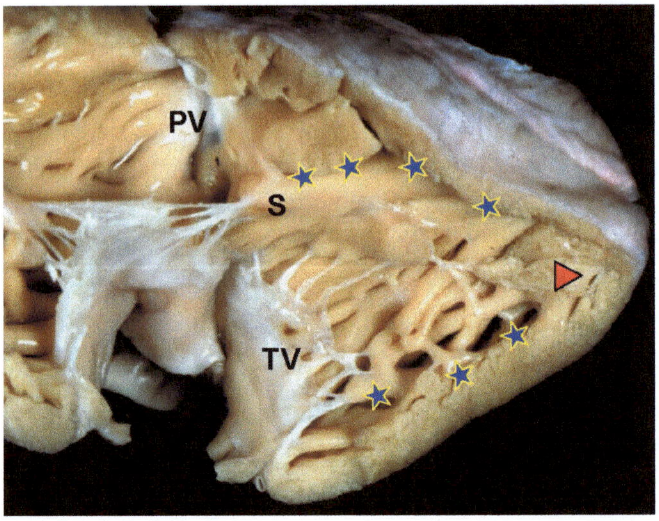

FIGURE 1.14. The right ventricle of a normal heart in which the parietal walls have been lifted away. The body of the septomarginal trabeculation is adherent to, or contiguous with, the septal surface. The triangle indicates a position that is analogous with the approximate position of the apex of the left ventricle. The extent of the ventricular septum, when perceived on the right side, is indicated by asterisks.

The ventricular cavity is usually, but not necessarily, dilated. Sometimes the walls are thin, but more often they are thicker than normal. Occasionally, gross hypertrophy is present. A quantitative analysis of the ages of our patients, the thicknesses of their right ventricular walls, and the inlet and outlet lengths of their right ventricles, showed that there were no statistical differences between all the four subgroups as categorised by the atresia or patency of their aortic and mitral valves. An evaluation of the histologic findings in the right ventricles of 59 hearts with hypoplastia of the left heart[25] has shown varying expressions of ischaemia, necrosis, fibrosis, calcification, and infarction amongst the different subgroups, but also without significant differences.

The normal morphologically right ventricle contains a prominent muscular component that overlies the ventricular septum and carries some of the tension apparatus supporting the tricuspid valve. This is the septomarginal trabeculation, which can be used as a marker against which to approximate both the extent of the cavity and the position of the apex of the left ventricle, also providing a surrogate for the length of the ventricular septum (Fig. 1.14). When we used this method to study our series of hearts with hypoplastic left ventricles, we found that, although the left ventricle may be small in these cases, for example with aortic atresia and either imperforate or patent but miniaturised mitral valve, its length is likely to be measureable, and correlates with the degree of hypoplasia of the septum as judged from the right ventricle. On the other hand, if a left ventricular cavity cannot be demonstrated by gross dissection, as frequently happens in cases with both aortic and mitral valvar atresia, it follows that a definitive septum cannot be identified. In these latter arrangements, the body of the septomarginal trabeculation, with its potential attachments to the septum, the septal leaflet of the tricuspid valve, and other tension apparatus, does not have a standard septal surface against which to lie. Consequently, there is increasing asymmetrical growth of the ventricles, with "remodelling" of the right ventricle (Figs. 1.15, 1.16 and 1.17).

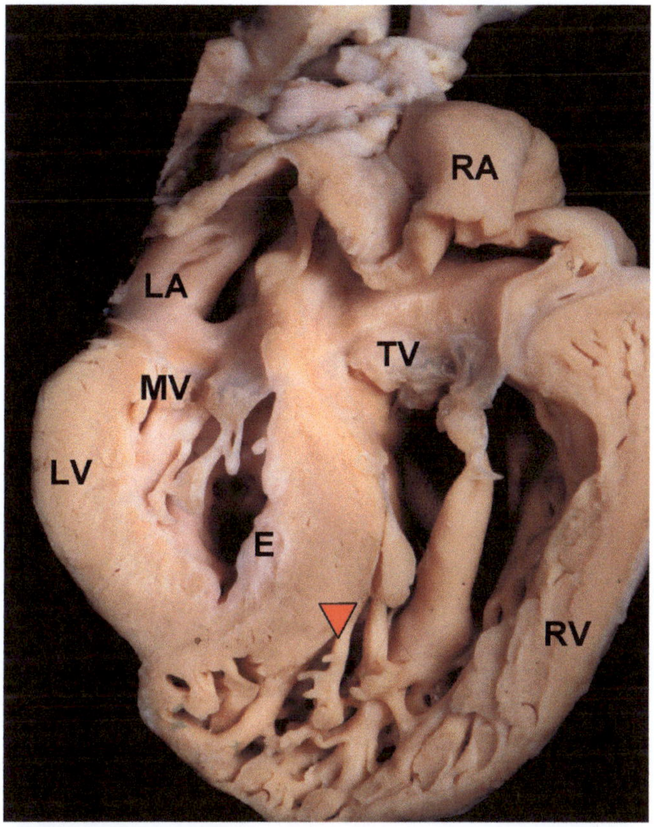

FIGURE 1.15. A view from the diaphragmatic surface showing all four cardiac chambers of a heart with aortic valvar atresia, patent but miniaturised mitral valve, and hypoplastic left ventricle with endocardial fibroelastosis. The anatomy is analogous with that of the heart shown in Figure 1.16. The triangle is placed approximately where the apex of the left ventricle would be perceived from the right side.

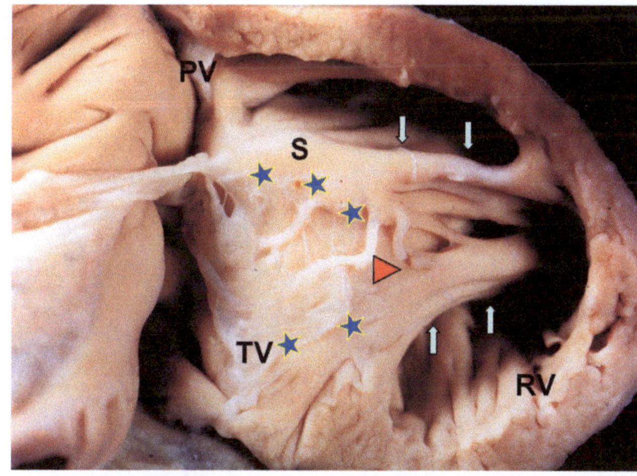

FIGURE 1.16. The septal surface of a thick-walled, dilated right ventricle in a heart with aortic valvar atresia and hypoplastic but patent mitral valve (see Figure 1.15). The septomarginal trabeculation is abnormal but is partially adherent to the hypoplastic septum. The extent of the ventricular septum is shown by asterisks. Sparse trabeculations stand freely (arrows), having become delaminated at a point that approximates the apex of the left ventricle (triangle).

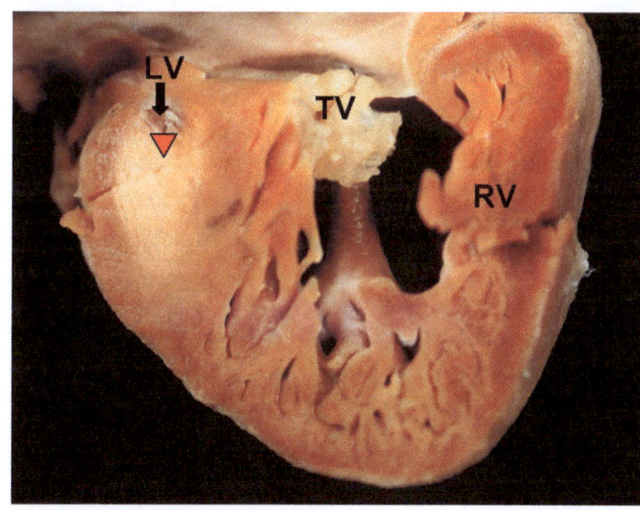

FIGURE 1.17. This heart has aortic valvar atresia and absent atrioventricular connection. It is viewed from its diaphragmatic surface. A left ventricular cavity is identifiable although extremely hypoplastic (arrow). The triangle represents its apex. The body of the septomarginal trabeculation (not seen) is adherent to the medial wall of the right ventricle, which is both hypertrophied and dilated.

We have used this term "remodelling" in these circumstances to describe the ensuing spectrum of variable, random attachments of the septomarginal trabeculation. In some of the hearts, the septomarginal trabeculation is contiguous with the medial or the superior wall of the right ventricle (Fig. 1.18). In others, it is free-standing (Fig. 1.19), or additionally rooted on the superior wall (Fig. 1.20). Significantly, when the septum is better formed, the remodelling seems to be proportionately less conspicuous (Fig. 1.16). Nonetheless, even in the latter cases, muscular trabeculations of the right ventricle can cross the cavity and produce another spectrum, that of abnormal anatomical spaces (Figs. 1.16, 1.19 and 1.20). If hypertrophied, the septomarginal trabeculation can rarely adhere to the superior wall of the right ventricle and obstruct the subpulmonary outflow tract (Fig. 1.21).

Within this anatomical spectrum, the cordal attachments of the anterior portion of the septal leaflet of the tricuspid valve may be tethered, as usual, to the septomarginal trabeculation, wheresoever it is adherent, but if it is an abnormal position, the cords may have become elongated. In addition, abnormal papillary muscular attachments are seen occasionally.[37] It seems to us that these morphological variations, observed in two-

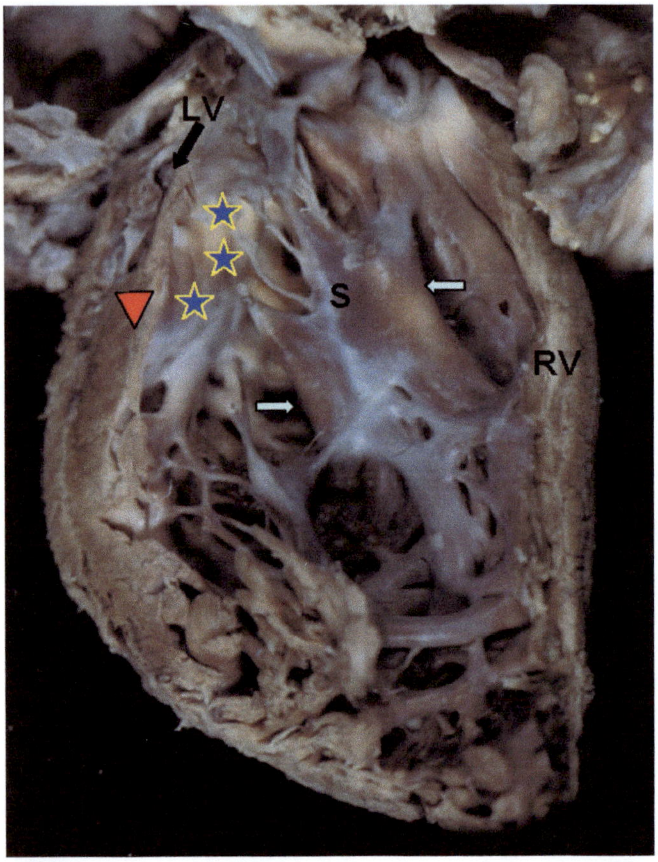

FIGURE 1.18. A view from the diaphragmatic surface of a heart with aortic valvar atresia and patent mitral valve. In contrast with the hearts shown in Figures 1.15, 1.16, 1.17 and 1.19, the right ventricle is thin-walled and dilated. The rudimentary cavity of the left ventricle is seen containing tension apparatus from the mitral valve. The extent of the hypoplastic septum is indicated as before (asterisks and triangle). The septomarginal trabeculation is contiguous with the superior wall of the right ventricle but probes will pass behind (arrows).

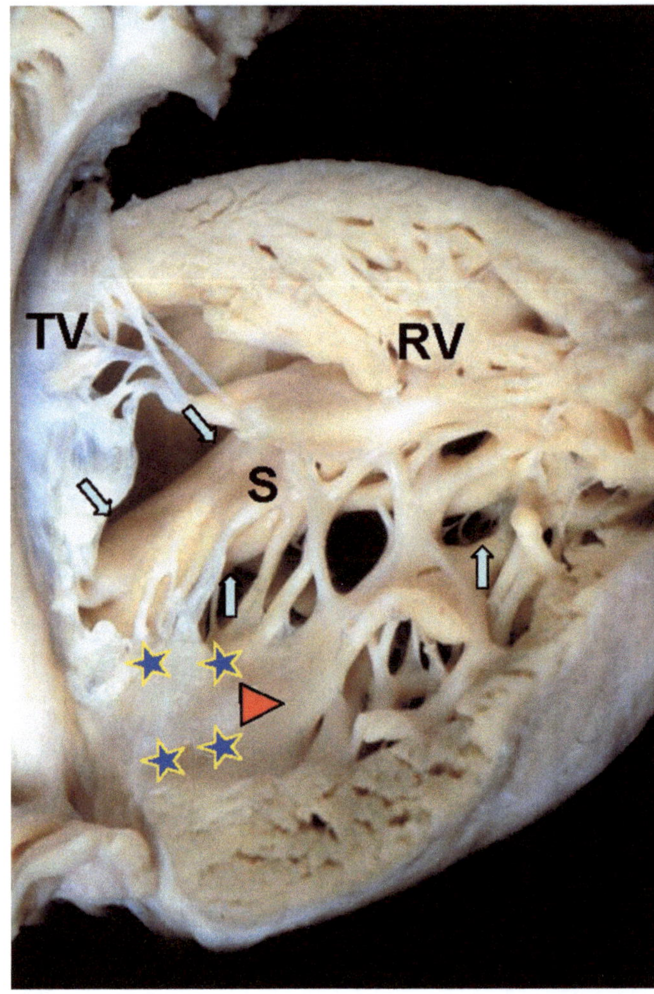

FIGURE 1.19. A slightly oblique, inferior view of the right ventricle in aortic valvar atresia with patent mitral valve. The left ventricle and septum (asterisks and triangle) are much smaller than as shown in Figures 1.15 and 1.16 but comparable with that shown in Figure 1.18. The body of the septomarginal trabeculation is almost completely delaminated and freestanding (arrows). A complex of other trabeculations produces anatomical "spaces" within the cavity of the ventricle.

fifths of our series, may have produced tricuspid valvar regurgitation during life.

We also noted that, in hearts in which the left ventricle was short but measureable, the hypoplastic septum bulged convexly into the right ventricle (Fig. 1.17). In the more exaggerated situation, the right side of the inlet septum wraps itself around the hypertrophied but smaller left ventricle, producing a pronounced curvature in the right ventricular cavity (Fig. 1.15). We found this type of morphology in one-third of our series with hypoplasia of the left heart. When the left ventricle was larger, however, the septum was seen to be straight, or even concave.[36]

With critical aortic stenosis, or imperforate or atretic aortic valve, or in the presence of a ventricular septal defect, both the ventricular cavities may run almost to the apex of the heart (Figs. 1.8, 1.10 and 1.22), with relatively normal measurements. These hearts may have thin left ventricular walls, with hypoplastic mitral valvar papillary muscles, and a straight ventricular septum. In this situation, the basal limbs and the body of the septomarginal trabeculation are usually adherent to the right side of the septum in the normal way. Only the apical trabeculations cross the ventricular cavity, although there may be abnormal muscle bars present. The anterior portion of the septal leaflet of the tricuspid valve is attached normally to the septomarginal trabeculation.

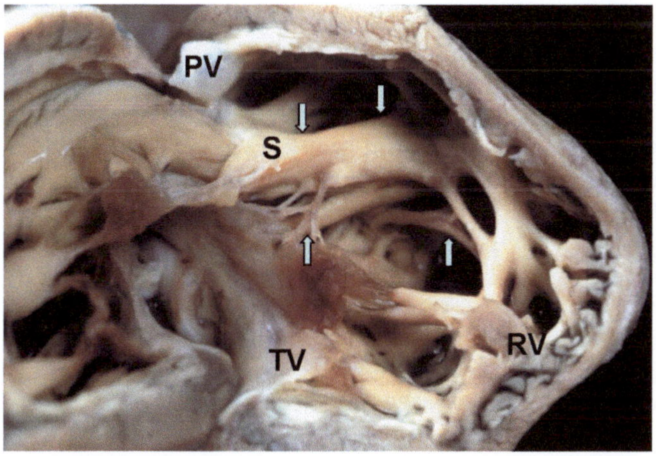

Figure 1.20. A left ventricular cavity is not identifiable on gross dissection in this heart with aortic valvar atresia and absent left atrioventricular connection. Thus, no identifiable ventricular septum is present. The right ventricle is thin-walled. The septomarginal trabeculation is standing freely (arrows), crossing toward the apical portion of the cavity with other robust trabeculations and producing anatomical spaces.

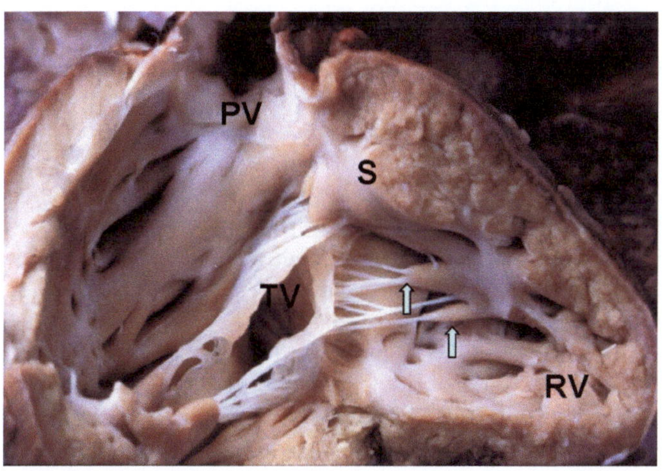

Figure 1.21. Neither a left ventricular cavity nor a ventricular septum is identifiable in this heart with aortic valvar atresia and absent left atrioventricular connection. In contrast with the heart shown in Figure 1.20, the right ventricle is hypertrophied. The body of the septomarginal trabeculation, also hypertrophied, has passed superiorly to become adherent to the superior wall of the right ventricle, thus producing anatomical subpulmonary obstruction. An inferior extension from the body of the septomarginal trabeculation is delaminated (arrows). Some of the cordal attachments to the tricuspid valve are long.

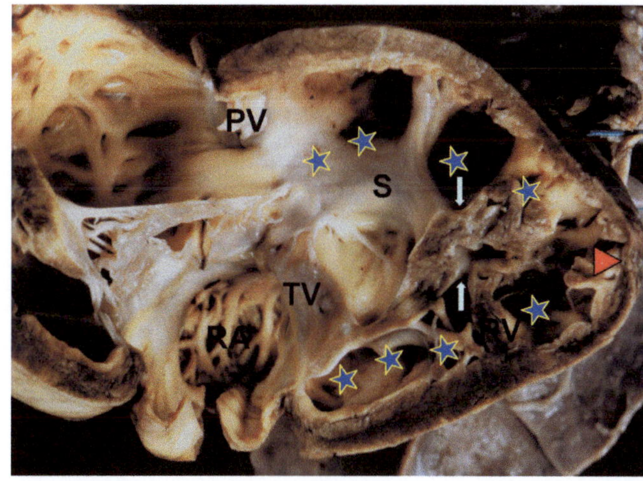

Figure 1.22. As a comparison, the right ventricle is thin-walled in this heart with critical aortic valvar stenosis. The body of the septomarginal trabeculation is adherent to the septum, becoming delaminated in the apical portion (arrows), where abnormal trabeculations cross the cavity. The left ventricle of this heart is also thin-walled but, although hypoplastic, has a normal length, as indicated (asterisks). The left side of the septum is shown in Figure 1.10.

COMMENTS

Following the original description by Lev,[39] and subsequent to the extensive study of Noonan and Nadas,[1] hearts with hypoplasia of the left ventricle have come to be well documented and extensively reviewed in the literature. When using the segmental approach,[3–5] description is straightforward. The pathological complex involving the left side of the heart, and the channels feeding the flow of blood to the systemic circulation, have been widely assessed. Perhaps surprisingly, since it is called upon to support the circulations after Norwood palliation, the right side of the heart, in particular the right ventricle, has received little attention. In this chapter, we have emphasised the features of the right side, hoping to redress this situation.

Our findings are also pertinent to questions that have been asked regarding surgical palliation. These address the issue as to whether certain anatomic patterns are less suitable for palliation than others, and in particular whether major problems are more likely to be found in the presence of a "blighted" left ventricle, or in the presence of fistulous communications with the coronary arteries. In this regard, Freedom's[22] opinion was that it would be preferable to palliate patients in whom the left ventricle and the mitral valve were virtually absent. In these hearts, he argued that the state of the left ventricular myocardium did not present the problems associated with high pressure or excessive perfusion of the coronary arteries, and by implication endocardial fibroelastosis, as was seen with patent mitral valve. Sauer and her co-workers[40] endorsed this view, finding that hypoplasia of the left heart with associated mitral atresia involved little pathology of the endocardium and the coronary arteries, and that this condition was the more favourable for palliative surgery. On the other hand, their study also demonstrated that endocardial fibroelastosis, and pathology of the coronary arteries, were not necessarily a constant feature in hypoplastic left ventricles with aortic atresia and patent mitral valve. We have not looked at the microanatomy and the histopathology of the coronary arteries, but we did observe that, on gross dissection, a small number of our cases with aortic atresia and patent mitral valve had not yet developed significant endocardial fibroelastosis. Baffa and her colleagues[25] have referred to the predisposition for endocardial fibrosis in those having hypoplasia of the left ventricle, pointing out that this is not the case in hypoplasia of the right heart, where there is a distinct association with coronary arterial fibroelastosis. Although evidence of coronary arterial abnormalities, and endocardial fibroelastosis, is greater in patients with hypoplasia of the mitral valve and aortic atresia, they found that there were no apparent differences in perfusion of the right ventricle amongst their anatomical subgroups. They concluded that, at that time, there was no justification for discriminating management on the basis of the mitral valve or the associated coronary arterial anatomy.[25] It also remains a fact that, as yet, there is no evidence from patients undergoing the Norwood procedure that survival is linked to the morphology of the left ventricle. On the other hand, it is well recognised that impediments to the pulmonary venous drainage may result in prenatal morbidity or mortality,[31] and that if pulmonary vascular disease is present at birth, this will negatively influence survival.[29] In this respect, there is support for the concept that palliative surgical procedures may benefit children who are awaiting heart transplantation.[30]

In contrast, the arrangement of the right ventricle in hypoplasia of the left heart could prove to be much more clinically significant. Amongst our series of hearts with aortic atresia and hypoplasia of the left ventricle and septum, in which the mitral valves were absent or imperforate, we recognised the extreme end of the spectrum of remodelling of the right ventricle. When the left ventricle could be identified in these hearts, it always showed absence of endocardial fibroelastosis. Next to these, along the spectrum, were hearts with aortic atresia and patent mitral valve, but with readily identifiable left ventricles and, usually, fibroelastosis. A pronounced convex curvature of the septum into the right ventricle was often present in this subgroup. Amongst all of our subsets, remodelling of the right ventricle appeared to be directly proportionate to the degree of hypoplasia of the septum. All of the other associated intracardiac and extracardiac abnormalities were independent factors. The associations between absence or the smaller size of the septum, and the greater degree of remodelling of the right ventricle, support the idea that the initial insult producing the remodelling had occurred early in the embryonic period. This suggests that early onset of disease is associated with the most severe end of the spectrum.[22] In cases in which the ventricular septum cannot be demonstrated by gross dissection, the insult almost certainly occurs prior to the process of delamination of the tricuspid valve. Remodelling of the musculature of the right ventricle, therefore, is to be anticipated. Nonetheless, as is well recognised by echocardiographers,[41,42] hypoplasia of the left heart may sometimes present late in gestation, when asymmetric growth of the ventricles is less exaggerated and remodelling of the right ventricle is not required. In terms of remodelling, separation of the septomarginal trabeculation from the right side of the septum, emphasising relatively large anatomic spaces behind it, has been discussed previously by Bharati and Lev.[43] Such spaces have sometimes been interpreted, on echocardiography, to be the definitive hypoplastic left ventricle.[44] Our observations endorse these concepts of the "freestanding" septomarginal trabeculation, together with the variable spaces to be seen in the right ventricle. We attribute this type of morphology to remodelling of the right ventricle. Some of the associated malformations include accessory papillary muscles[43] that occasionally show a "pallisade-like" morphology. "Mitralisation" of the tricuspid valve, also reported by Bharati and Lev,[43] was identified in our material as a bifoliate valve supported by only two papillary muscles. Stamm and colleagues[37] also described the bifoliate arrangement of the right atrioventricular valve associated with significant dysplasia of the muscular and tendinous supporting structures. Although there is evidence of a wide variety of subtle abnormalities of the right ventricle in hypoplasia of the left heart, there are no features that are constant. Thus, there is diversity between the morphological features of the right ventricles of individual hearts, even though the hearts may be similar when categorised by their left-sided morphology. These differences almost certainly explain the tricuspid valvar dysfunction and regurgitation[36,45] known to affect some, but not all, patients in the clinical setting. Although anatomical variations in the ascending aorta and the aortic arch have major implications for the operative repair of coarctation, permitting the avoidance of homograft material,[21] they hold few implications for the standard Norwood operation.[46] Anatomical variations in the right ventricular mural thickness, the septomarginal trabeculation, the tricuspid valve and its tension apparatus, and also of the right ventricular outflow tract, in contrast, almost certainly do have implications for right ventricular function. In this respect, the recent work of Altmann and his co-authors[47] showed that intermediate and overall survival after the Norwood operation for hypoplasia of the left heart was significantly decreased in their patients with initially diminished right ventricular function. Actuarial survival 18 months after the Norwood procedure was 93% for their patients with initially normal right ventricular function, whereas it was no more than 47% for those with abnormal right ventricular function prior to surgical palliation. Subtle differences in the morphology of the right ventricles of patients with hypoplasia of the left heart, therefore, almost certainly have the potential to influence the long-term outcome of the Norwood operation. As yet, unfortunately, we cannot use our morphologic findings to predict clinical outcomes.

References

1. Noonan JA, Nadas AS. The hypoplastic left heart: an analysis of 101 cases. Pediatr Clin North Am 1958;5:1029–1057.
2. Tchervenkov CI, Tahta SA, Jutras LC, Bjorkland MJ. Biventricular repair in neonates with hypoplastic left heart complex. Ann Thorac Surg 1998;66:1350–1357.
3. Anderson RH, Becker AE, Freedom RM, et al. Sequential segmental analysis of congenital heart disease. Pediatr Cardiol 1984;5:281–288.
4. Van Praagh R. The segmental approach to diagnosis in congenital heart disease. In: Birth Defects: Original Article Series. Baltimore: Williams and Wilkins, 1972; 8:4–23.
5. de la Cruz MV, Nadal-Ginard B. Rules for diagnosis of visceral situs, truncoconal morphologies and ventricular inversions. Am Heart J 1972;84:19–32.
6. de la Cruz MV, Anselmi G, Angel Romero BS, Monroy G. A qualitative and quantitative study of the ventricles and great vessels of normal children. Am Heart J 1960; 46:675–690.
7. Eckner FAO, Brown BW, Davidson DL, Glagor S. Dimensions of normal hearts after standard fixation by controlled pressure coronary perfusion. Arch Pathol 1969;88:497–504.
8. Elzenga NJ, Gittenberger-de Groot AC, Oppenheimer Dekker A. Coarctation and other obstructive aortic arch anomalies: their relationship to the ductus arteriosus. Int J Cardiol 1986;13:289–308.
9. van der Horst RL, Hastreiter AR, DuBrow IW, Eckner FA. Pathologic measurements in aortic atresia. Am Heart J 1983;106:1411–1455.
10. Robert Arnold, personal communication, May, 2003.
11. Pellegrino A, Deverall PB, Anderson RH, et al. Aortic coarctation in the first three months of life. An anatomicopathological study with respect to treatment. J Thorac Cardiovasc Surg 1985;89:121–127.
12. Ho SY, Anderson RH. Coarctation, tubular hypoplasia and the ductus arteriosus. Histological study of 35 specimens. Br Heart J 1979;41:268–274.
13. Elzenga NJ, Gittenberger-de Groot AC. Coarctation and related aortic arch anomalies in hypoplastic left heart syndrome. Int J Cardiol 1985;8:379–389.
14. von Reuden TJ, Knight L, Moller JH, Edwards JE. Coarctation of the aorta associated with aortic valvular atresia. Circulation 1975;3:951–954.
15. Hawkins JA, Doty DB. Aortic atresia: morphologic characteristics affecting survival and operative palliation. J Thorac Cardiovasc Surg 1984;88:620–626.
16. Machii M, Becker AE. Nature of coarctation in hypoplastic left heart syndrome. Ann Thorac Surg 1995; 59:1491–1494.
17. Becker AE. Segmental aortic hypoplasia or how to interpret the flow concept. Int J Cardiol 1988;20:247–255.
18. Jonas RA, Lang P, Hansen D, Hickey P, Castaneda AR. First stage palliation of hypoplastic left heart syndrome. The importance of coarctation and the shunt size. J Thorac Cardiovasc Surg 1986;92:6–13.
19. Van Meurs-Van Woezik H, Krediet P. Measurements of the descending aorta in infants and children: comparison with other dimensions. J Anat 1982;135:273–279.
20. Moulaert AJ, Bruins CC, Oppenheimer-Dekker A. Anomalies of the aortic arch and ventricular septal defects. Circulation 1976;53:1011–1015.
21. Karl TR, Sano S, Brawn WB, Mee RBB. Repair of hypoplastic or interrupted aortic arch via sternotomy. J Thorac Cardiovasc Surg 1992;104:688–695
22. Freedom RM. Atresia or hypoplasia of the left atrioventricular and/or ventriculo-arterial junction. In: Anderson RH, Macartney FT, Shinebourne EA, Tynan M, eds. Paediatric Cardiology, Vol 2. Edinburgh, London, Melbourne, New York: Churchill Livingstone, 1987 pp 737–765.
23. Lloyd TR, Evans TC, Marvin WJ Jr. Morphologic determinants of coronary blood flow in the hypoplastic left heart syndrome. Am Heart J 1986;112:666–671.
24. O'Connor WN, Cash JB, Cattril CM, Johnson GL, Noonan JA. Ventriculocoronary connections in hypoplastic left hearts. An autopsy microscopic study. Circulation 1982;66:1078–1086.
25. Baffa JM, Chen SL, Guttenberg ME, Norwood WI, Weinberg PM. Coronary artery abnormalities and right ventricular histology in hypoplastic left heart syndrome. J Am Coll Cardiol 1992;20:350–358.
26. Raghib G, Bloemendaal RD, KanjuhVI, Edwards JE. Aortic atresia and premature closure of foramen ovale myocardial sinusoids and coronary arteriovenous fistula serving as outflow channels. Am Heart J 1965;70:476–480.
27. Buhlmeyer K, Simon B, Mocellin R, Sauer U. Clinical angiocardographic and functional studies in the assessment of critical valvular aortic stenosis. In: Godman MJ,

Marquis RM, eds. Paediatric Cardiology. Heart Disease in the Newborn, Vol 2. Edinburgh: Churchill Livingstone, 1979;220–235.

28. Pinto CA, Ho SY, Redington A, Shinebourne EA, Anderson RH. Morphological features of the levoatriocardinal (or pulmonary-to-systemic collateral) vein. Pediatr Pathol 1993;13:751–761.
29. Brann S, Brawn WJ, Raafat F, Sreeram N. Neonatal pulmonary vascular disease in hypoplastic left heart syndrome. Ann Thorac Surg 1995;60:433–434.
30. Collins MH, Darragh RK, Caldwell RL, Turrentine MW, Brown JW. Short-term survivors of paediatric heart transplantation: an autopsy study of their pulmonary vascular disease. J Heart Lung Transplant 1995;14:1116–1125.
31. Rychik J, Rome JJ, Collins MH, DeCampli WM, Spray TL. The hypoplastic left heart syndrome with intact atrial septum: atrial morphology, pulmonary vascular histopathology and outcome. J Am Coll Cardiol 1999; 34:554–560.
32. Vasquez JC, Rabah R, Delius RE, Walters HL. Hypoplastic left heart syndrome with intact atrial septum associated with deletion of the short arm of chromosome 18. Cardiovasc Pathol 2003;12:102–104.
33. Trowitzsch E, Bano-Rodrigo A, Burger BM, Colan SD, Sanders SP. Two-dimensional echocardiographic findings in double orifice mitral valve. J Am Coll Cardiol 1985;6:383–387.
34. Lopez W. Aortic atresia without significant hypoplasia of the ascending aorta. AJR 1964;92:888.
35. Gittenberger-de Groot AC, Wenink ACG. Mitral atresia: morphological details. Br Heart J 1984;51:252–258.
36. Reyes A 2nd, Bove EL, Mosca RS, Kulik TJ, Ludomirsky A. Tricuspid valve repair in children with hypoplastic left heart syndrome during staged surgical reconstruction. Circulation 1997;96:II-341–343; discussion II-344–345.
37. Stamm C, Anderson RH, Ho SY. The morphologically tricuspid valve in hypoplastic left heart syndrome. Eur J Cardiothorac Surg 1997;12:587–592.
38. Bharati S, Nordenberg A, Brock RR, Lev M. Hypoplastic left heart syndrome with dysplastic pulmonary valve with stenosis. Pediatr Cardiol 1984;5:127–130.
39. Lev M. Pathologic anatomy and interrelationship of the aortic tract complexes. Lab Invest 1952;1:61–70.
40. Sauer U, Gittenberger-de Groot AC, Geishauser M, Babic R, Buhlmeyer K. Coronary arteries in hypoplastic left heart syndrome. Histopathologic and histometrical studies and implications for surgery. Circulation 1989; 80:I-168–I-176.
41. Allan LD, Cook A, Sullivan I, Sharland GK. Hypoplastic left heart syndrome: effects of fetal echocardiograph on birth prevalence. Lancet 1991;337:959–961.
42. Sharland GK, Chita SK, Anderson RH, Tynan M, Cook AC, Allan LD. Left ventricular dysfunction in the fetus: relation to aortic valve anomalies and endocardial fibroelastosis. Br Heart J 1991;66:419–424.
43. Bharati S, Lev M. The surgical anatomy of hypoplasia of aortic tract complex. J Thorac Cardiovasc Surg 1984;88:97–101.
44. Ruschhaupt DG, Moshiree M, Lev M, Bharati S. Echocardiogram in mitral-aortic atresia. False identification of the ventricular septum and left ventricle. Pediatr Cardiol 1980;1:281–285.
45. Barber G, Helton JG, Aglira BA, et al. The significance of tricuspid regurgitation in hypoplastic left heart syndrome. Am Heart J 1988;116:1563–1567.
46. Norwood WI, Lang P, Casteneda AR, Campbell DN. Experience with operations for hypoplastic left heart syndrome. J Thorac Cardiovasc Surg 1981;82:511–519.
47. Altmann K, Printz BF, Solowiejczyk DE, Gersony WM, Quaegebeur J, Apfel HD. Two-dimensional echocardiographic assessment of right ventricular function as a predictor of outcome in hypoplastic left heart syndrome. Am J Cardiol 2000;86:964–968.

Chapter 2

The Phenotype During Human Fetal Development

Andrew C. Cook

Over the years, a number of potential etiologies for the hypoplastic left heart syndrome have been suggested, many of which refer to processes thought to have occurred during early fetal development. Some of these relate to abnormal embryogenesis, some to abnormal fetal blood flow, and still others to fetal infection.[1–14] If any of these theories is correct, it should accurately reflect the abnormal structure of the fetal heart, as seen in humans with left heart hypoplasia. In fact, closer inspection shows that most of the conclusions from previous studies were, of necessity, speculative, being based either on the study of cardiac anatomy as seen later in life in human neonates, with extrapolation back to processes thought to occur during embryonic or fetal life, or else on experiments utilising mechanical obstruction to blood flow through the developing heart in birds and animals.[9,10,15–17] It is only recently, with the advent of fetal echocardiography, that the gross structure of the human fetal heart has been able to be seen in the setting of hypoplasia of the left heart.[18,19] As a result of prenatal diagnosis, it is also possible to look in detail at the structure of the developing heart with left-sided hypoplasia in pregnancies that have been interrupted, and determine whether, for instance, theories relating to abnormal flow or infection are likely to be correct. Furthermore, we can also obtain a much wider view of the overall malformation itself, from the late first trimester of development to term, since this population will be relatively unselected. The information reported in this chapter is based on study of a large number of fetal hearts with varying forms of left-sided hypoplasia, examined anatomically, histologically, and with respect to potential etiologic factors.

HOW CAN WE DEFINE HYPOPLASIA OF THE LEFT HEART DURING DEVELOPMENT?

Defining a left heart that is hypoplastic during fetal life, at a time when growth and development is not yet complete, brings its own specific problems. Although the main structure of the heart has been built by the 8th week of gestation, growth of the chambers and arteries is still dramatic from this point until birth. Consequently, it is theoretically relatively easy for hypoplasia of the left heart to be produced, even late in gestation, simply by arrest of normal growth. Although such "passive" hypoplasia has been produced experimentally in animals, it appears to occur relatively infrequently in humans during very late gestation.[15] Furthermore, as pregnancy progresses, there is a dramatic increase in variance in the overall size of normal fetuses, and therefore of their hearts. In other words, while all fetuses may be of similar size, and have similarly sized left ventricles at 10 weeks' gestation, by birth one fetus may have grown significantly more than another, and in absolute terms its heart and left ventricle will be significantly larger. This means that definitions based on absolute size are impossible. To overcome this, it is possible to interpret data from abnormal fetuses in terms of deviation from the mean, or z-scores.[20] This was the approach taken in my study. In turn, such interpretation requires careful modelling of how both the mean and standard deviation change in the normal situation with gestational age.[21–23] Even utilising such techniques, the measurements taken are valid only for the gestational age at which they are taken, and tell nothing about subsequent growth of the left heart in an individual fetus. For the anatomist, there is no way around this dilemma,

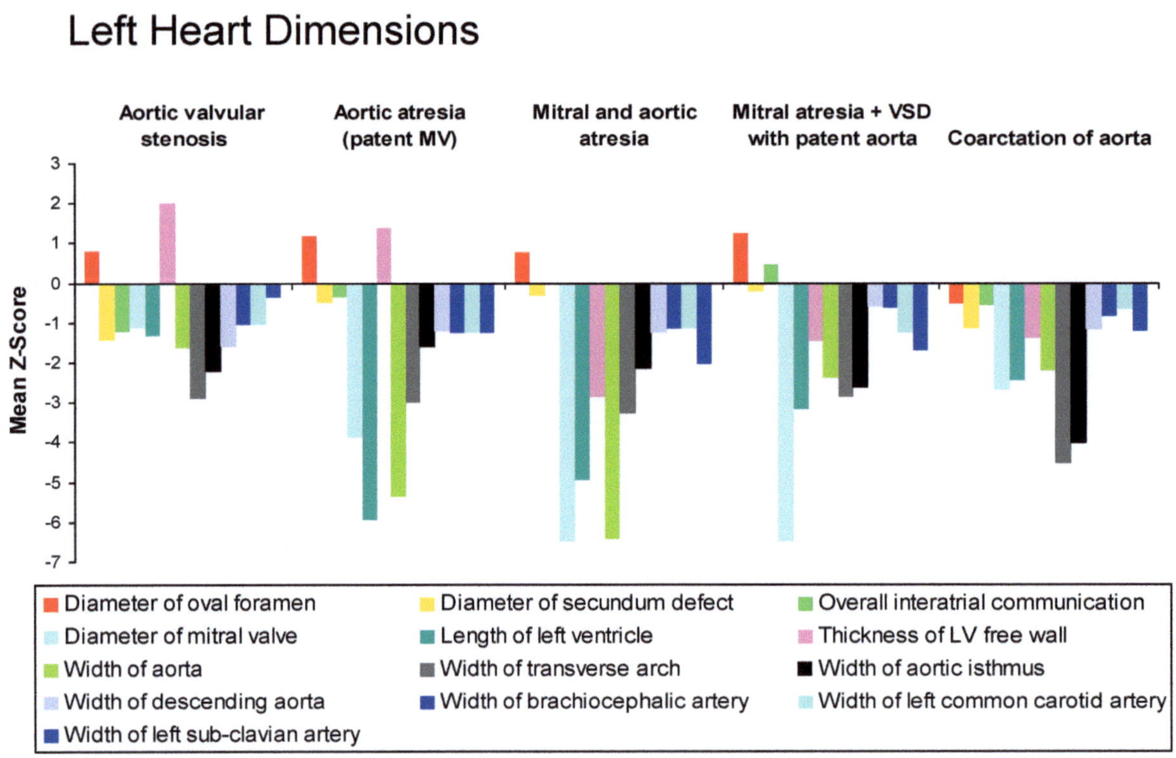

A

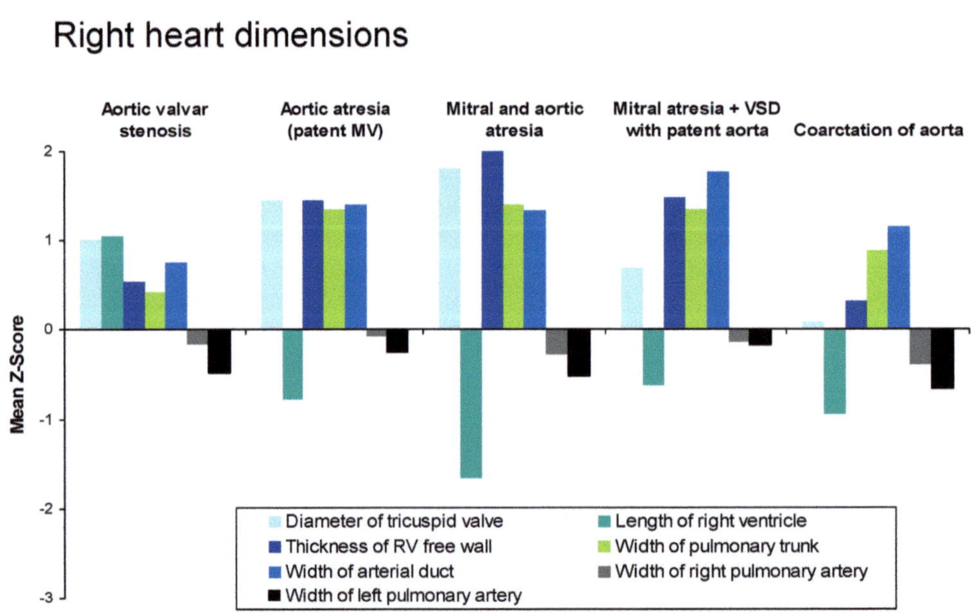

B

FIGURE 2.1. These graphs show mean z-scores for the cardiac structures across five forms of left heart abnormality. They illustrate the degree of deviation from the mean found in normal fetuses. A: Left heart structures, most of which are smaller than the mean. B: The corresponding structures in the right heart. Most are increased, compared to normal, showing that the right heart has already begun to adapt. Nonetheless, the increase is not as marked as the degree of hypoplasia of left heart structures. LV, left ventricle; RV, right ventricle; MV, mitral valve; VSD, ventricular septal defect.

since all measurements are of necessity taken at a single point in time. For measurements taken during life, it is possible to construct individual and longitudinal velocities of growth for each fetus, based on serial measurements of the components of the left heart, and to compare these growth velocities to those occurring in normal fetuses.[23] So far the utility of such complex comparisons has not been proven for fetuses with left heart hypoplasia.

A Solitary Malformation or Heterogeneous Group of Malformations?

The above discussion would be unnecessary if left heart hypoplasia in the fetus were a uniform disease. Experience from studying morphology in neonatal life has clearly shown that the hypoplasia can exist in a number of anatomical forms, but it has become apparent that in the fetus still others need to be considered. As in the neonate, there are some instances where the left ventricle is virtually nonexistent. In these instances, there is no problem in definition. So, for the fetus with mitral and aortic atresia and an intact ventricular septum, for instance, there is no difficulty in saying the left ventricle is hypoplastic, since it lacks both its inlet and outlet. Such hearts form a major part of the developmental study outlined below. Also providing little problem with definition are those fetuses with mitral atresia, a ventricular septal defect, and a patent aortic root, or those with aortic valvar atresia. In all examples of these malformations in the current study, the dimensions of the left heart vary somewhat from fetus to fetus, but in mid-gestation fall well below normal values (Fig. 2.1A). More problematic are those fetuses with patent mitral or aortic valves associated with either an abnormality of the aortic arch or stenosis of the aortic valve. As can be seen from Figure 2.1A, left ventricular dimensions in these hearts are closer to the mean, and therefore there is greater difficulty in predicting left ventricular growth with age. These malformations have been included in this study since it is known from clinical experience that some will have progressed to develop left heart hypoplasia by birth.[24–26] It is these more "borderline" fetuses, therefore, that potentially are the missing links between normality and left heart hypoplasia. It should be noted at this stage that this list of candidate hypoplastic left hearts is not exhaustive, but simply an example to show that the heterogeneity of the phenotype that exists during fetal life is more widespread than that seen following birth. Small left ventricles can also be associated, for instance, with common arterial trunk, tetralogy of Fallot, or double outlet right ventricle, but these are generally rare associations, even during fetal life.[27] More frequently, degrees of left heart hypoplasia can also accompany an atrioventricular septal defect.[28] Although this type of malformation is not considered further in this chapter, since deficiency of the atrioventricular structures makes them fundamentally different, it is worth bearing in mind that factors that limit left ventricular growth in this situation may be similar to those in fetuses with more classical forms of left heart hypoplasia.

A "Syndrome" or Malformation Sequence?

As described in Chapter 4 in this book, it is debatable whether or not the term "hypoplastic left heart syndrome" is appropriate, since most reports do not describe extracardiac malformations as part of the condition, let alone a consistent pattern of malformations constituting a syndrome. Nevertheless, the presence or absence of extracardiac malformations may well depend not only on the age group being studied, but also on the manner in which the subjects are sought. In this developmental study, to define the incidence of these anomalies, the external and internal extracardiac anatomy of the fetuses was examined in detail at autopsy. It is of note, then, that extracardiac malformations were found in between one-tenth and one-third of fetuses with left heart hypoplasia, following exclusion of those with a known chromosomal abnormality (Fig. 2.2). In general, the malformations were various and random, and so do not indicate a specific "syndrome". The exception to this rule was in three of the 223 fetuses, all of whom had a scimitar malformation and hypoplasia of the right lung, together with mitral and aortic atresia and a hypoplastic left heart (Fig. 2.3). These cases may well represent a previously unrecognised syndrome.

If not a syndrome, in the strict sense of the word, then what does hypoplasia of the left heart represent? Is it that the various lesions, such as atresia or stenosis of the mitral valve, hypoplasia of the aortic root, and coarctation of the aorta follow from a single cardiac insult, or are they separate lesions that just happen to coexist? In other words, are they a "malformation sequence" within the heart? To answer these and other questions, I have concentrated on the incidence and relationship between four components of the heart in fetuses with varying forms of left heart hypoplasia to see if there is a common pattern between 5 of the different anatomical forms of the malformation.

MORPHOLOGY OF THE ATRIAL SEPTUM

Patency of the atrial septum, and the nature of the flow of blood from the inferior caval vein to the left atrium, has received widespread attention in past studies.[9,10]

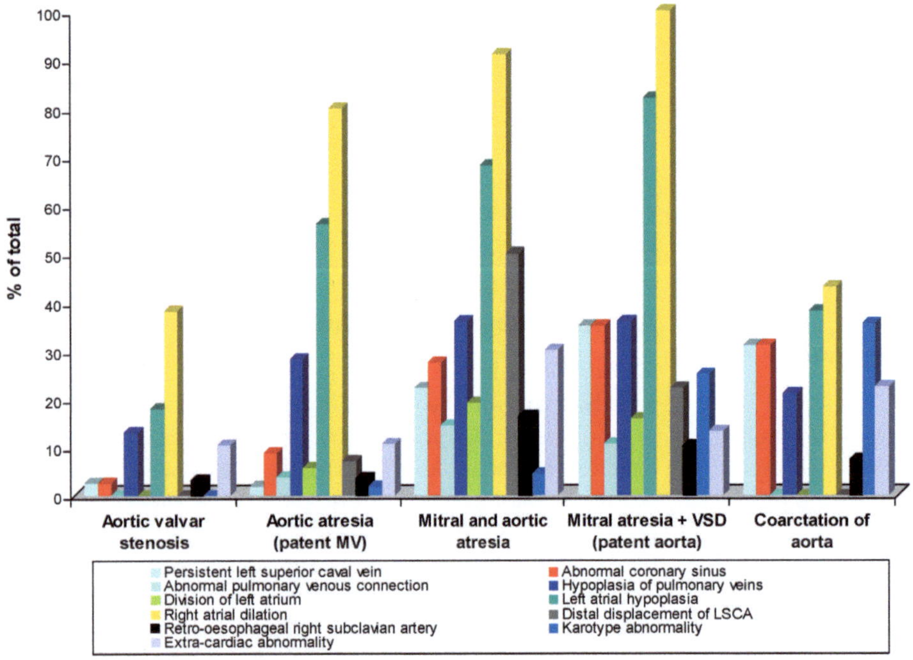

FIGURE 2.2. This graph shows the incidence of both intracardiac and extracardiac anomalies in various forms of left heart hypoplasia. The frequency of associated anomalies is greater for severe forms of left heart hypoplasia and less for those with less marked hypoplasia. This fits with the hypothesis that severe forms have their genesis early in embryonic development whereas less severe forms may be acquired following formation of the heart. LSCA, left subclavian artery; MV, mitral valve; VSD, ventricular septal defect.

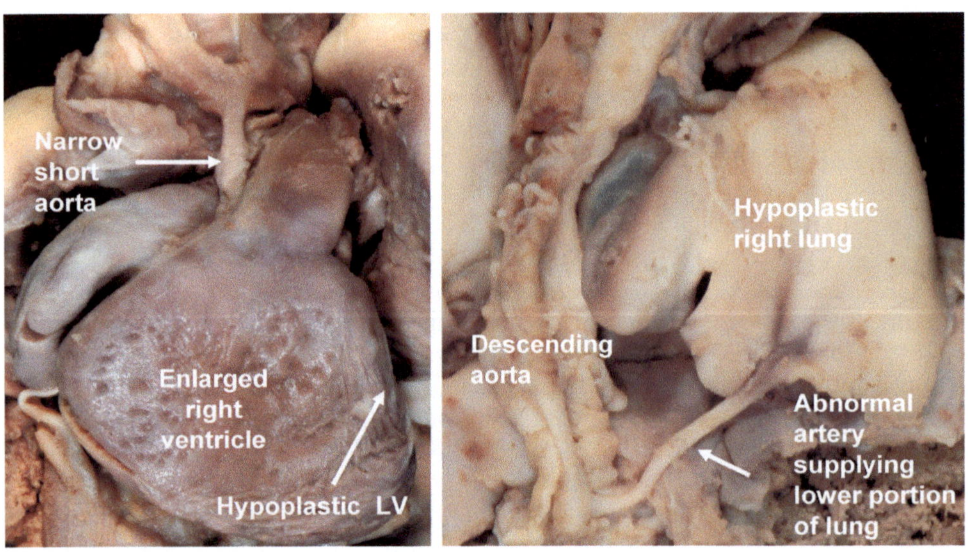

A B

FIGURE 2.3. An example of one of the fetuses with severe hypoplasia of the left heart and scimitar malformation. There was mitral and aortic atresia (A), the right lung was half the expected size, and the both pulmonary venous and arterial supply to the lower lobe was abnormal (B). LV, left ventricle.

Indeed, hypoplastic left heart syndrome is described, in many texts, as a "flow-related" disease.[29] It has been proposed that "premature closure" of the atrial septum occurs during fetal life in patients with left heart hypoplasia, and that this effectively "blocks" the passage of blood from inferior caval vein to the left-sided structures, which atrophy as a result.[9] Support for this hypothesis comes, in principle, from experiments performed in late gestation in the fetal lamb and the chick, in which the communication between the two atria is blocked, artificially, leading subsequently to degrees of left ventricular hypoplasia.[15,16] The main evidence in humans, however, is much less convincing. Studies have shown abnormalities in the structure of the atrial septum, such as premature closure, an abnormal relationship between the atrial septum and left atrium, and hypoplasia of the eustachian valve, which normally directs blood into the left atrium via the oval foramen.[9,10] Undoubtedly, these abnormalities all occur, and are also to be found, with varying frequency, in fetuses with left heart hypoplasia (Fig. 2.4). Previously, however, these anomalies have been interpreted as the cause of the left-sided abnormality. There is good evidence, from the phenotype of fetal hearts together with

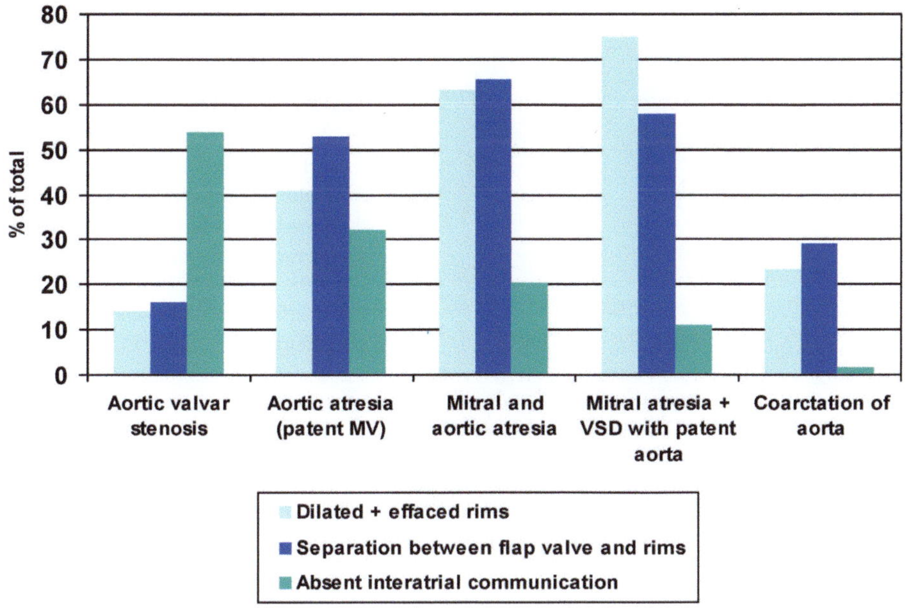

FIGURE 2.4. This graph shows the trend in abnormalities of the interatrial communication across five groups of fetuses with varying forms of hypoplasia of the left heart. The frequency of dilatation, effacement, and separation between components increases with severity of hypoplasia of the left heart. In contrast, the frequency of a closed interatrial communication decreases.

clinical information, that, apart from one important exception, the anomalies of the atrial septum are more likely to be secondary to preexisting disease in the left heart. The problem in interpretation has arisen, I believe, because previous investigators were looking at the fully formed neonatal heart, and extrapolating backward to fetal development.

When analysed in the mid-trimester of pregnancy, three things are clear regarding the structure of the atrial septum. First, abnormalities of the atrial septum are frequent and varied in fetuses with left heart hypoplasia. There can be enlargement of the interatrial communication, with effacement and dilation of its borders, and separation between its component parts. Alternatively, there is sometimes unequivocal evidence of premature closure, or restriction, of a previously well-formed interatrial communication. Then, there can also be evidence of primary absence or hypoplasia of the secondary defect, which should normally develop at approximately 4 to 5 weeks' gestation within the flap valve guarding the oval foramen. The variety seen suggests the existence of more than a single mechanism. Both dilation and premature closure of the atrial septum are likely secondary features readily explainable as a result of preexisting left heart disease. On the other hand, failure of formation, or delay in perforation, of the atrial septum suggests a primary problem in embryonic development, and warrants further experimental investigation.

Second, a trend in these anomalies exists amongst the differing forms of the disease, which suggests different mechanisms to those previously proposed. Overall, fetuses with mitral and aortic atresia have an enlarged

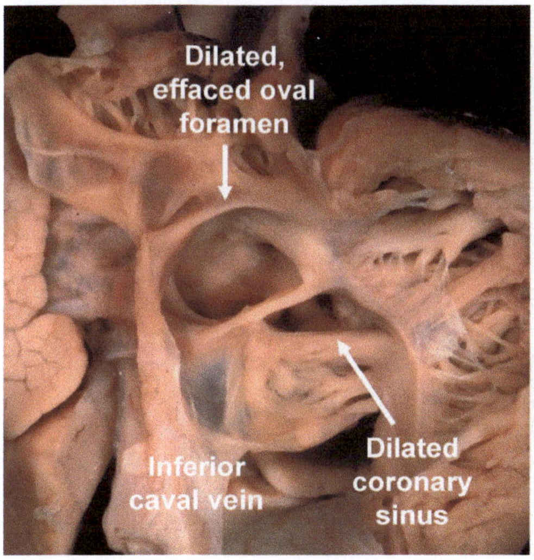

A

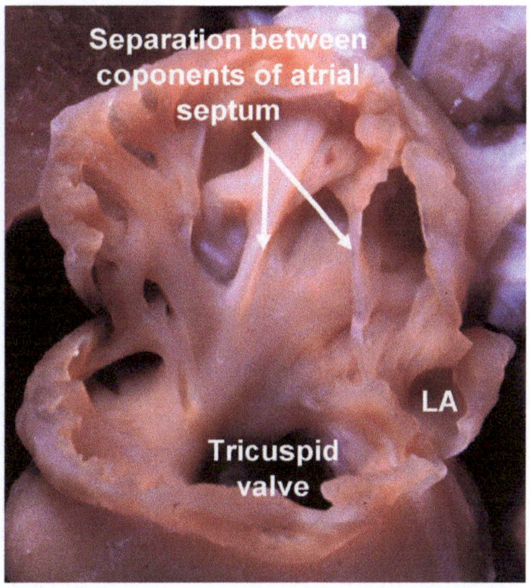

B

FIGURE 2.5. Anomalies of the interatrial communication seen most frequently in fetuses with mitral and aortic atresia. The rims of the oval fossa are dilated and enlarged (A), and there is separation between the rim of the oval fossa and the flap valve, which is thin and membranous (B). Both fetuses are 20 weeks' gestation. LA, left atrium.

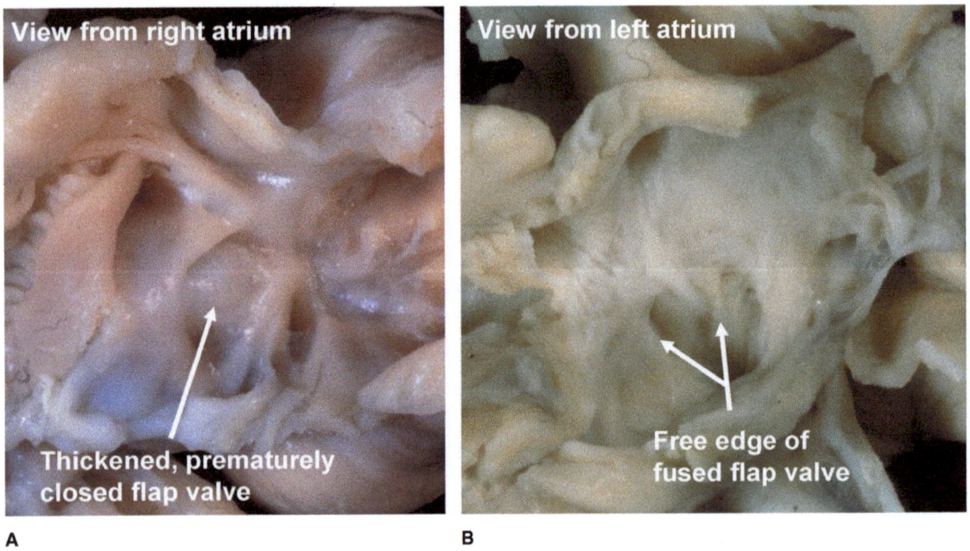

FIGURE 2.6. Anatomy of the atrial septum seen most frequently in fetuses with aortic valvular stenosis. The flap valve is prematurely thickened and muscularised and bulges toward the right atrium. When viewed from the left, the free edge of the thickened flap valve can be seen fused to the rims of the oval foramen. This suggests that the flap valve was initially perforate but subsequently closed prematurely.

and effaced oval foramen (Fig. 2.5). Since these fetuses are also known to have bidirectional but predominantly left-to-right flow across the atrial septum during life, we see these abnormalities of the atrial septum as secondary to this abnormal flow, and as a result of preexisting left heart hypoplasia.[30] At the other end of the spectrum, fetuses with aortic valvar stenosis, overall, have the smallest interatrial communications due to premature closure of the atrial septum (Fig. 2.6). This, I believe, is explainable on the basis of left ventricular dysfunction, mitral regurgitation leading to increased left atrial volume and pressure, and sealing of the flap valve against the atrial septum, as occurs in the normal fetus following birth.[31] Again, therefore, this form of abnormality is likely to be the result of preexisting left ventricular dysfunction, or an abnormality of the aortic valve. Of note, the dimensions of the left heart in this group of fetuses are the largest amongst all forms of left heart hypoplasia. In consequence, secondary closure of the atrial septum, as I have proposed, may well have led to subsequent left-sided hypoplasia had pregnancy continued. Such progression from a near-normal sized but dysfunctional left ventricle to hypoplasia of the left heart has been documented clinically. If these two ends of the spectrum are correct, then one might expect fetuses with aortic valvar atresia to show characteristics of both groups, since in these fetuses there is obstruction to both inlet and outlet from the left ventricle. This is exactly what is found on examination of fetal hearts. As a result, the interatrial communication in these fetuses as a group is of near-normal size, since in some it is small and in others enlarged (Fig. 2.7).

The final type of atrial septal anatomy, primary agenesis or hypoplasia of the secondary interatrial communication, is perhaps the most interesting in terms of etiology. It is seen most frequently amongst fetuses with anomalies of the aortic arch such as coarctation and interruption, but it is also evident in occasional fetuses amongst other forms of left-sided hypoplasia, for instance those with mitral and aortic atresia. As discussed, during development, perforation of the primary atrial septum is essential, following cardiac septation, so

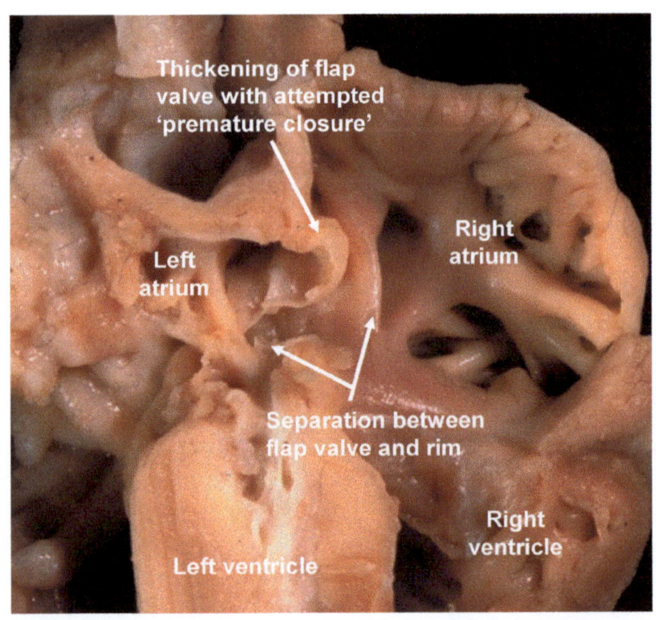

FIGURE 2.7. In fetuses with aortic valvular atresia, a mixture of morphology, illustrated in Figures 2.5–2.6 is seen. In this particular fetal heart, shown in four-chamber section, the superior rims of the oval foramen are somewhat effaced and lean toward the mouth of the superior caval vein within the right atrium. There is separation between the flap valve and the rims. Nonetheless, the flap valve itself is thickened and bulging toward the right in an attempt to close against the rims.

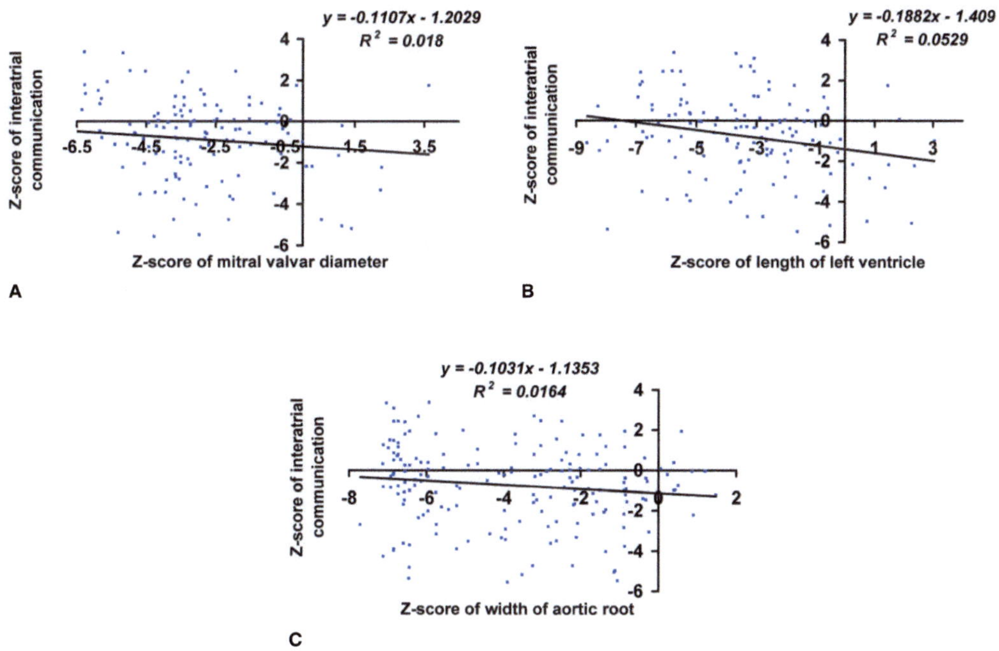

FIGURE 2.8. These three graphs illustrate that there is poor correlation between z-scores for the interatrial communication and those for the length of the mitral valve (A), left ventricular length (B), and the aortic root (C).

that blood can flow from the inferior caval vein to the left heart. This finding implies a possible etiologic link between delay or failure of perforation of the primary atrial septum and development of the left heart. That having been said, the left heart can clearly grow in the absence of an interatrial communication, since in some fetuses with imperforate atrial septums, left heart dimensions still fell on the third centile for age.

Third, there is no correlation between the overall size of the interatrial communication and the size of the mitral valve, the left ventricle, or the aortic root (Fig. 2.8). This would be surprising if left heart hypoplasia resulted from anomalies of the atrial septum. Instead, it suggests that the state of the atrial septum is independent of development of the left heart, at least during the stages of fetal development studied.

THE NATURE OF THE LEFT ATRIOVENTRICULAR AND VENTRICULO-ARTERIAL JUNCTIONS AND VALVES

The fact that makes the lack of correlation between the morphology of the atrial septum and left heart dimensions even more convincing is that there is much better correlation between the sizes of the left heart structures "downstream" of the atrial septum. In other words, the size of the mitral valve, when patent, correlates well with the size of the left ventricle and with the dimensions of the aortic root (Fig. 2.9). This rule appears to be general, since it applies not only to severe forms of the disease such as aortic atresia with patent mitral valve, but also to more borderline cases such as coarctation of the aorta, or aortic valvar stenosis. It means that the constellation of abnormalities found in left heart hypoplasia could well represent a "malformation sequence" within the heart, starting from a single point. Whether the starting point for this sequence is inadequate expansion of the left atrioventricular junction, or inadequate expansion of the left ventricle, or the outflow, remains to be determined experimentally. Evidence presented below suggests that the starting point is likely to differ in the various forms of the disease, but also that more than a single mechanism could operate in some forms of the disease.

Thus, there has long been a debate about the nature of the left atrioventricular junction in mitral atresia. Some authors have proposed that, while there is no patent connection between the left atrial and the left ventricular cavities, there is a fibrous connection or tract between the two chambers, suggesting that the cavities may have been in continuity at some stage of development.[32] Others have argued convincingly that there are only two forms of mitral atresia, namely those seen in the setting of a right atrium connected to a morphologically right ventricle with an imperforate mitral valve, and those with complete absence of the left atrioventricular connection.[33,34] They argue that fibrous tracts are simply remnants of the atrioventricular septum, continuous with the central fibrous body, and do not represent an atrophied left atrioventricular junction or mitral valve. How can these two points of view be resolved? Both of these studies are based on neonates. It is quite

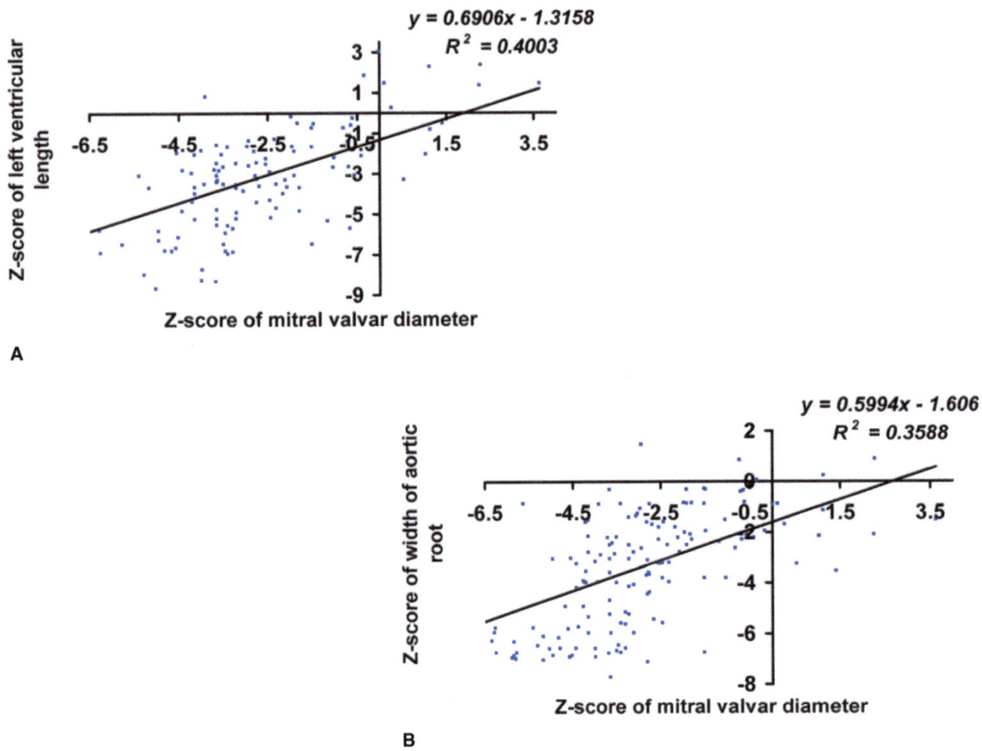

FIGURE 2.9. In contrast, these graphs illustrate the better correlation that exists between left heart structures downstream of the atrial septum. Z-scores of the mitral valve correlate well with those of both the left ventricle (A), and the aortic root (B).

possible, therefore, that fetal specimens might provide further clues to the mechanism of production of mitral atresia, since they are much closer to the point of genesis of the malformation itself. When I performed such an examination, I found only complete absence of the left-sided atrioventricular connection, or an imperforate mitral valve, with no evidence of fibrous tracts.[35] Of note, however, was that the incidence of an imperforate mitral valve is much higher than has been reported postnatally, being present in almost one-half of fetuses with mitral atresia.[34–37]

Could these imperforate valves atrophy and fibrose with age, accounting for at least some of the fibrous tracts noted by others? It is possible, in fetuses, to see with the microscope that the imperforate membrane points to the inlet of the left ventricle, rather than its outlet, and also to see tiny tendinous cords and papillary muscles. The dimension of the imperforate valve and its tension apparatus is also truly minute, measuring less than 1 mm in diameter at the extreme end (Fig. 2.10). Consequently, with rapid growth of the heart between 21 weeks and birth, it is quite feasible that the nature of this potential connection between the left atrium and ventricle could be lost, and that the imperforate valve could atrophy, becoming a fibrous strand continuous with the atrioventricular membranous septum. Should there then also be overgrowth, or hypertrophy, of the left atrium, the connection between the fibrous strand and left atrium would also be lost. The mode of atresia would then appear, to all intents and purposes, as an absent connection. There is good evidence that such overgrowth can occur, even in the middle trimester of pregnancy, in fetuses with left atrial hypertrophy, due to

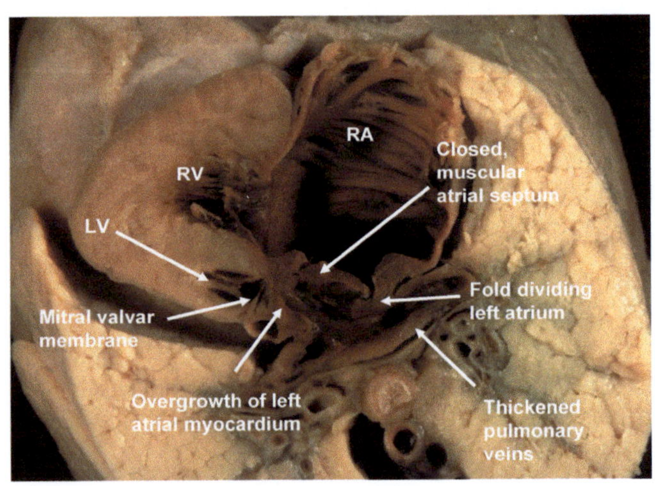

FIGURE 2.10. This section through a fetal heart with an imperforate mitral valve illustrates how the mode of atresia can become obscured during fetal life. The interatrial communication is closed in this fetus, leading to thickening of the left atrium and overgrowth of the mitral valve by atrial musculature. When viewed from the left atrium it would appear as though the mitral valve has never developed, but when viewed from the left ventricle, the membrane together with its supporting tension apparatus can still be seen. RA, right atrium; RV, right ventricle; LV, left ventricle.

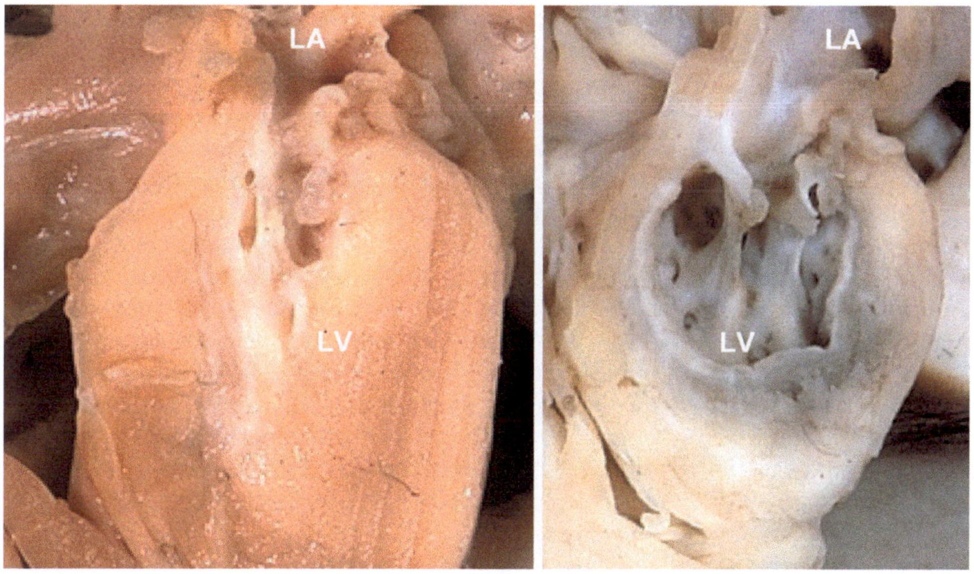

A B

FIGURE 2.11. These are two examples of the mitral valve in fetuses with aortic atresia and intact ventricular septum. Although they are both formed and patent, they are abnormal in structure. The valve shown in A is minuscule and of similar size to the imperforate mitral valve shown in Figure 2.10. The leaflets, tendinous cords, and papillary muscles in both examples are arranged like a funnel leading into the left ventricle. LA, left atrium; LV, left ventricle.

lack of an interatrial communication. In these fetuses, even relatively large imperforate mitral valves can become overgrown with atrial myocardium (Fig. 2.10).

Of note, the size of the imperforate mitral valve and tension apparatus seen in some fetuses with mitral atresia is very similar to the size of the patent mitral valve seen at the severe end of the spectrum in fetuses that only have aortic valvar atresia. In these, the valve is patent but abnormal, and together with its tension apparatus consists of a single sheet of tissue arranged like a tiny funnel (Fig. 2.11). Overall, this suggests to me that there is a complete spectrum in formation and inadequate expansion of the left atrioventricular junction and development of the mitral valve, from complete absence, to miniature imperforate valves, to miniature funnel-like but patent valves.

The same type of analysis can then be applied to the left ventriculo-arterial junction in fetuses with aortic atresia to provide clues to the mechanisms of maldevelopment. In the mid-trimester of pregnancy, there are major differences in the mode of atresia, depending on

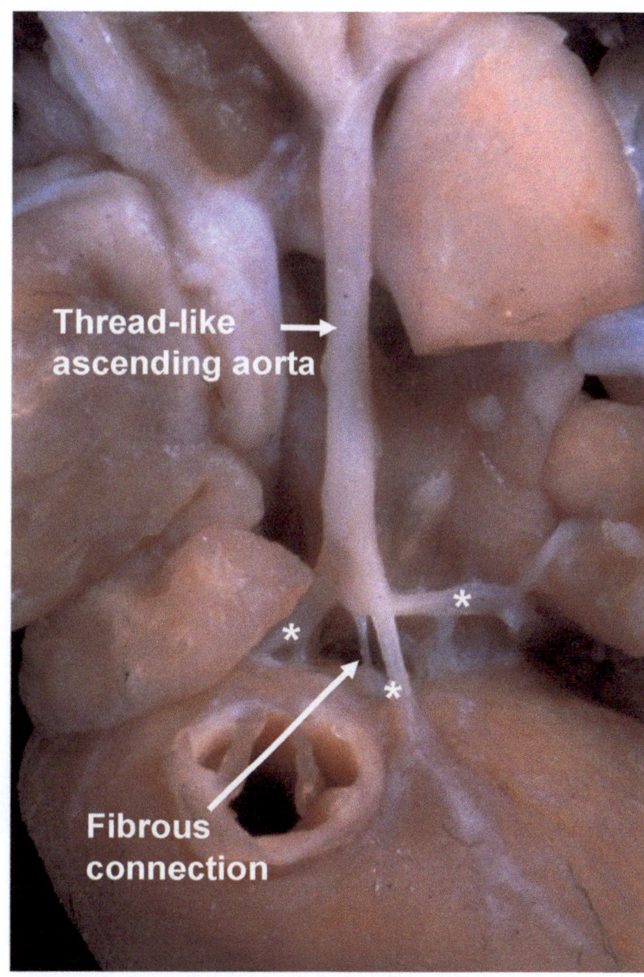

FIGURE 2.12. This close-up shows the nature of the aortic root found in the majority of fetuses with mitral and aortic atresia. The ascending aorta is less than 1 mm in external width. It ends in a shallow "bulb," which gives rise to the coronary arteries (asterisk). The base of the bulb is connected to the ventricular myocardium by a very narrow fibrous strand, which on sectioning connects with the fibrous core of the heart. It suggests that, in these fetuses, the connection between left ventricle and aorta has never developed.

whether the mitral valve is formed and patent, or not. Again, with age, many of these clues appear to be lost, as they are less evident in fetuses around the time of birth. When there is a "full house" of both mitral and aortic atresia, then in four-fifths of fetuses, there will be fibromuscular tissue obstructing the route from the left ventricle to the aortic valve. From the arterial side, the aortic root, which itself is consistently tiny, ends blindly in an extremely shallow "bulb," but still gives rise to the coronary arteries at the point where the bulb joins the tubular portion of the aorta. The bulb itself retains only a very narrow fibrous connection with the central fibrous body of the heart (Fig. 2.12). Thus, it seems that the connection between left ventricle and aorta has been abnormal from the outset in these cases, perhaps even since the time of division of the arterial trunk into pulmonary and aortic vessels. It would not seem feasible easily to reconstitute a pathway between left ventricle and aorta in these fetuses. The same anatomy is also seen in one-quarter of fetuses with aortic atresia and a patent mitral valve. In the majority of these latter cases, nonetheless, the left ventricular outflow tract is patent, and leads to an imperforate aortic valvar membrane. The aortic root is significantly larger in these fetuses, and more variable in size, reaching near normal proportions in some instances (Fig. 2.13).

This morphology raises the possibility that there was, at one stage of development, a patent connection between the left ventricular outflow tract and aortic root. In those fetuses with larger aortic roots and left ventricles, it also raises the possibility of restoring the connection between aorta and left ventricle by perforation of the thickened imperforate valvar membrane. There is then further evidence that, in three-quarters of fetuses with aortic atresia and a patent mitral valve, the connection between left ventricle and aortic root was initially patent. Detailed inspection and sectioning of the imperforate aortic valve from its arterial side shows that there are three small sinuses, separated by three ridges or raphe, in hearts with an imperforate aortic valve. That much has been known from the time of the original descriptions of aortic atresia by Abbott and others (Fig. 2.14).

Closer examination, however, demonstrates that the size of the raphe is not always equal. Indeed, in the majority of hearts, the ridge between the noncoronary and left coronary sinuses is higher than the others (Fig. 2.14). Inspection of the left ventricular outflow tract from below shows a corresponding pattern. Usually three, but sometimes one or two, small triangular spaces can be seen spaced around the circumference of the imperforate aortic valve, each sitting beneath one

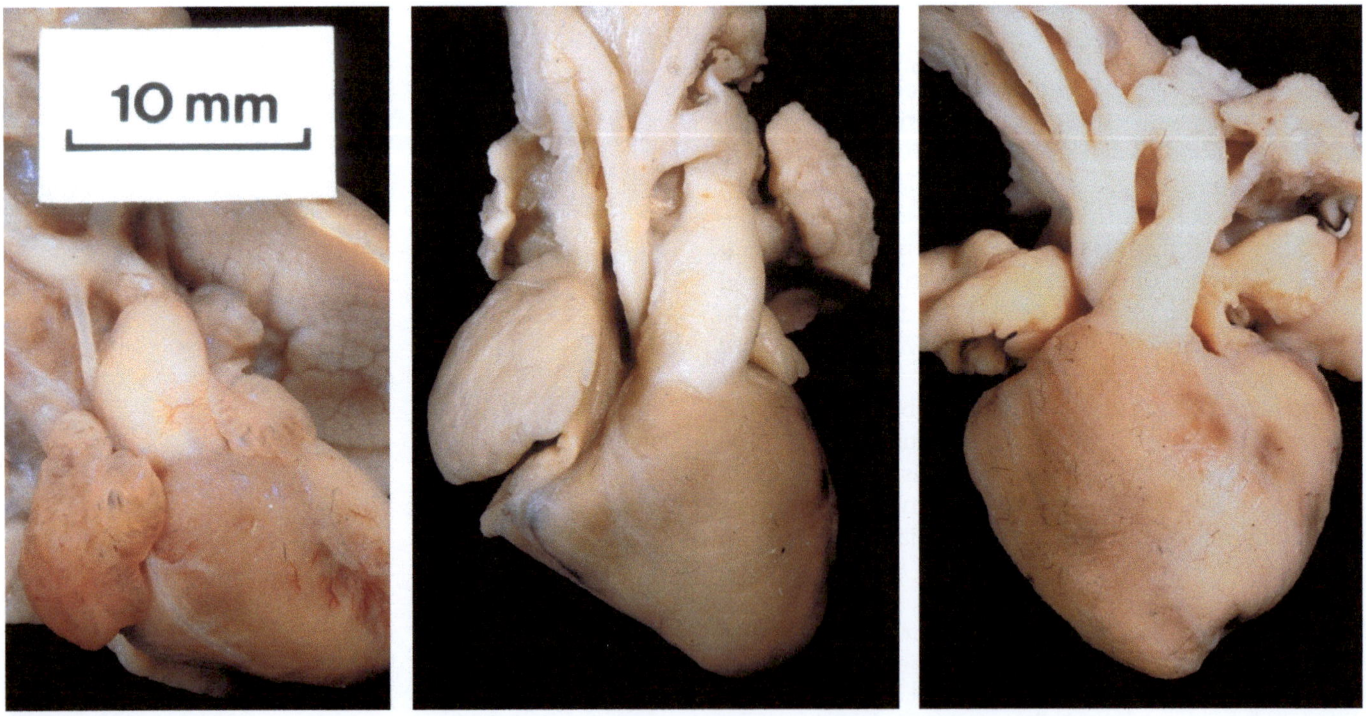

A B C

FIGURE 2.13. In contrast to Figure 2.12, the ascending aorta in fetuses with aortic atresia and a patent mitral valve can be of variable size, in fetuses of the same gestational age. It ranges from less than 1 mm (A) through an intermediate stage (B), to a near-normal size (C), and indicates that, in this setting, atresia can occur at differing stages of development.

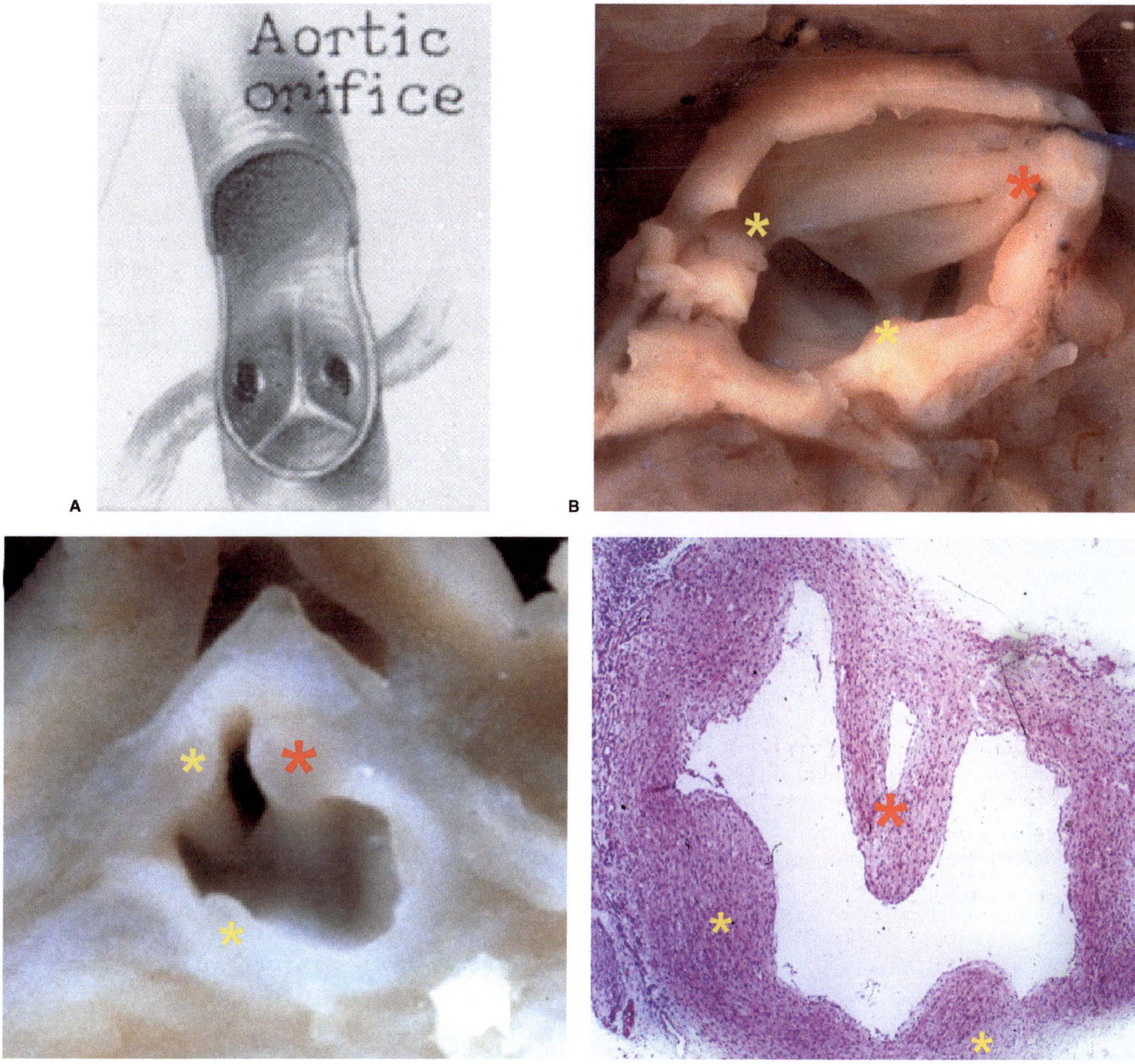

FIGURE 2.14. The nature of the imperforate aortic valve found in most fetuses with aortic atresia and a patent mitral valve. A: Part of an illustration from the atlas of Maude Abbott. It shows three symmetrical raphes (asterisk) radiating from the centre of the aortic root, but in fetuses there is often asymmetry between the raphes with the one between the left and noncoronary sinuses being larger than the others (red asterisk in Figure 2.14b). On sectioning this raphe can be seen as an arch rather than a solid structure (D). This mirrors the asymmetric arrangement seen in fetuses with aortic valvular stenosis (B), in which the remaining pin-hole orifice is also situated between the left and noncoronary sinuses. The morphology suggests that in most fetuses with aortic atresia the aortic valve was patent but stenotic.

of the raphes found in the base of the aortic root. These spaces seem to be remnants of the interleaflet triangles seen in the normal heart, which are usually of equal size, and rise between and to the full height of each sinus. In the setting of an imperforate aortic valve, the triangles are much thinner and shorter than normal. Usually, the triangle between the left and noncoronary sinus is larger than the remainder. This, then, corresponds with the size of the raphe within the aortic root, creating an asymmetric imperforate aortic valve with a "hollow" raphe between the non- and left coronary sinuses (Fig. 2.15). Comparing this structure with the arrangement seen in aortic valvar stenosis shows some clear similarities. In valvar stenosis, there is usually extensive

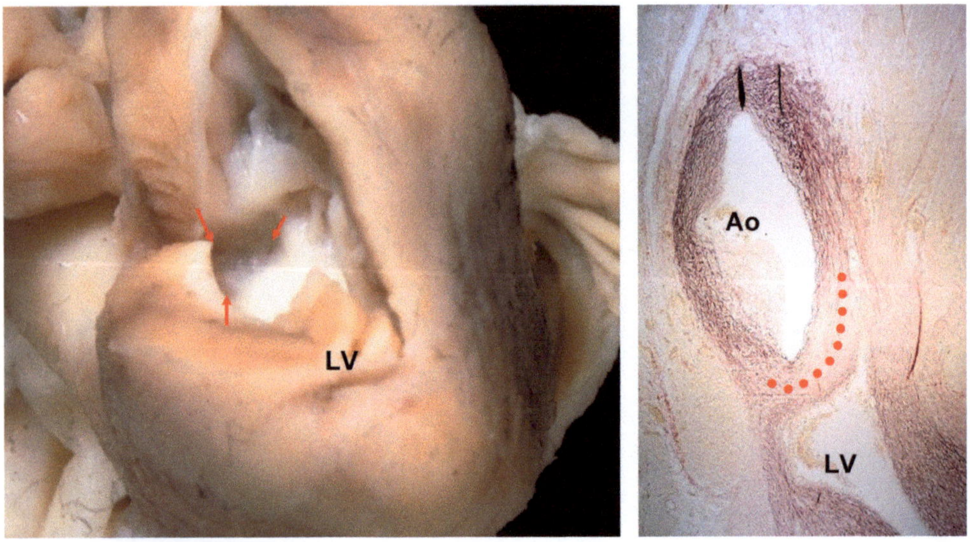

A B

FIGURE 2.15. In this fetus with aortic atresia and a patent mitral valve, the imperforate aortic valve is viewed from below, from the left ventricle (A) and has been sectioned lengthways (B). A: The arrows point to the remnants of three interleaflet triangles, which surround the imperforate valve. B: Sectioned across one of the triangular remnants, the pathway between the left ventricle and aortic root is seen blocked by an imperforate membrane (dotted line). LV, left ventricle; Ao, Aortic root.

fusion between the leaflets of the aortic valve, with corresponding hypoplasia of the triangles of space between the fused leaflets. Usually, however, the leaflets between the left and noncoronary sinuses are unfused, and the interleaflet triangle between them is well formed. This gives the stenotic aortic valve an asymmetric shape, with a keyhole orifice situated between the sinuses of the left and noncoronary leaflets (Fig. 2.14). The similarity is clear between the structure of this valve and that of the imperforate aortic valve seen in fetuses with aortic atresia. It suggests that progression from severe aortic stenosis to atresia could have occurred in most fetuses with a patent mitral valve and that there is initially a potential but likely stenotic connection between the aortic root and the left ventricle. The cause for the initial stenosis at valvar level remains to be determined.

These, then, are just examples of the overlap in phenotype seen in some forms of left heart hypoplasia. As mentioned at the outset, the left ventricle can also be small in hearts with abnormalities of the aortic arch, such as coarctation of the aorta or interruption. In neonates with these conditions, both the mitral and aortic valves are frequently abnormal in structure, although their size is frequently normal. It is no surprise, therefore, that abnormalities are also common in the middle trimester fetus. In particular, the frequency of abnormal aortic valves in the fetus is remarkably similar to that found in neonates, and approximates to one-half. What is different is that the severity of the aortic valvar abnormality is greater in the fetus than in the neonate with coarctation. Although it is feasible that both the left ventricular hypoplasia and the aortic valvar abnormalities could recover in these fetuses from fetal to neonatal life, this remains to be proven. An alternative explanation would be that the fetal and neonatal groups are different populations, the fetuses being the severe end of the spectrum of coarctation with consequent poorer natural survival. This hypothesis is supported by other differences between the two age groups. For instance, in the fetal group, females predominated, with a ratio of female to males of 2 to 1. In neonates, there are equal ratios between genders, but in infants with coarctation, males are reported to be more common.[38,39] Overall, this suggests that female fetuses with coarctation may have poorer survival, since they are more severely affected.

THE VENTRICULAR MYOCARDIUM AND CORONARY ARTERIES

Two features of the ventricular myocardium are important for survival in infants with left heart hypoplasia. As mentioned previously, the ability of the left ventricle to act, or not to act, as a pump for the systemic circulation is crucial in terms of clinical and surgical management. Some would propose that this also should be the manner in which the "syndrome" or "complex" should itself be defined.[40] In functional terms, when unable to support the systemic circulation, the anatomy of the left heart becomes virtually irrelevant, since it is excluded from the circulation. In terms of etiology, however, the phenotype of the left ventricle could still provide important clues. The structure and function of the right ventricle is also of crucial importance, particularly if it is required to support the systemic circulation via palliative procedures such as the Norwood series of operations. An abnormality of the coronary arterial supply to the right ventricle is one parameter that can profoundly affect its function. Some other aspects of the right ventricle have already been illustrated in Chapter 1, but it is unclear when these adaptations take place. For the fetus, appropriate growth, and perhaps even survival, will depend on the response of the right ventricle to the change in haemodynamic load that will occur as a result of left-sided hypoplasia. Does the right ventricle enlarge to cope with the increased haemodynamic load, or does it respond by other mechanisms? When does this take place during development, and is the response always sufficient for fetal survival and, perhaps more importantly, to enable normal fetal development to take place? As we will see, answers to some of these questions are available from study of the fetal phenotype. As always, they open as many avenues for research as they close.

First, with regard to major anomalies of the coronary arteries, there has been some debate about the presence and significance of coronary arterial malformations in patients with hypoplasia of the left heart.[41,42] These anomalies would be significant, if present, since they could impact on the coronary arterial supply to the right ventricle, which, following palliative surgery, becomes the systemic ventricle. Overall, coronary arterial anomalies have not been found in patients with mitral atresia, but have been documented angiographically, by ultrasound and histologically, in those with aortic atresia or stenosis and a patent mitral valve.[41–43] The time of onset of these anomalies, and their incidence, has always been rather uncertain. From my histological studies of fetal hearts, I have been able to determine that coronary arterial abnormalities are already present in one-third of fetuses with aortic atresia or stenosis and a patent mitral valve, but are not found in any other form of the disease. The anomalies consist of ventriculocoronary communications, but thickening of the coronary arterial walls is also present, even early in the second trimester of pregnancy.[35] Significantly, in two of the five abnormal cases, a coronary artery supplying the right ventricle was also affected. This questions whether right ventricular function would have become compromised (Fig. 2.16). These findings confirm that coronary arterial anomalies are present from early in development. The pattern follows similar rules to that seen in hypoplasia of the right heart, in that they are more prevalent the smaller the affected ventricle.[35,44,45] They differ in that the affected coronaries are usually only mildly dilated and tortuous. The communications are often narrowed, and sometimes occluded on their ventricular side by endocardial fibroelastosis. This is not the case in hypoplasia of the right heart, in which there is little endocardial fibroelastosis, and in which the communications are often large and unobstructed.[44] In may well be, then, that the natural history of these abnormal ventriculocoronary communications is different in hypoplasia of the right versus hypoplasia of the left heart. In hypoplasia of the left heart, the evidence suggests that the channels may exist early in fetal life, but become narrowed or occluded with time by fibrosis at their junction with the ventricle.

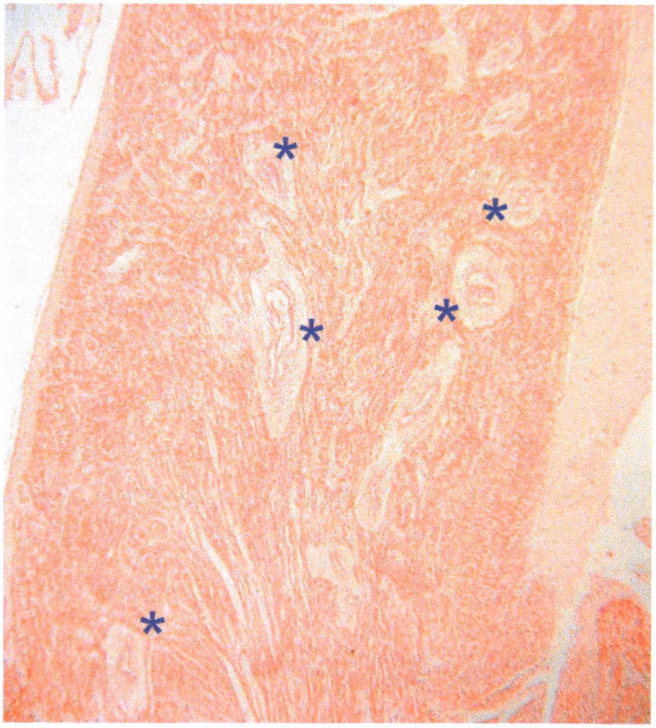

FIGURE 2.16. This histologic section is taken through the anterior papillary muscle of a 21-week gestation fetus with aortic atresia and patent mitral valve. A ventriculocoronary communication was traced to the right ventricle in this fetus and the coronary arteries in the papillary muscle are abnormal and thick-walled.

Second, with respect to the morphology of the abnormal left ventricle, there is generally good correlation between the size and thickness of the left ventricle and the dimensions of its inlet and outlet. As has been proven from studies in neonates, the left ventricle is generally narrow and slit-like in the presence of both mitral and aortic atresia, but round and thick-walled in the presence of a patent mitral valve and aortic atresia.[36] Measurements in fetuses show that the free wall is thinner than normal in fetuses with mitral atresia, but normal in structure. In the presence of mitral atresia and a ventricular septal defect, the ventricular free wall is of nearer normal thickness. Similarly, in most fetuses with aortic valvar atresia the free wall will be thicker than normal with increased fibrosis and active endocardial fibroelastosis, but in those with severe hypoplasia of the mitral valve, the free wall is thinner than in the structurally normal fetal heart, and shows less severe fibrotic changes. Overall, these measurements indicate that the expansion of the left ventricle is critically dependent on the amount of blood flowing into the chamber, whether it be via the mitral valve or via defects in the ventricular septum. This rule can be applied to instances of borderline hypoplasia, as found in fetuses with aortic coarctation. Fibrotic changes develop only in the presence of an obstructed outlet, a patent inlet, and an intact ventricular septum. In fetuses with aortic valvar stenosis, a third change is apparent over and above thickening and fibrosis of the left ventricle. In these fetuses, there is frequently conspicuous calcification of the left ventricular myocardium. This is independent of changes in the coronary arteries, the degree of endocardial fibroelastosis, and the pathology of the aortic valve (Fig. 2.17). It cannot easily be explained on the basis of myocardial ischaemia, as has been suggested from previous postnatal studies.[46,47] Instead, it is much more likely to be a primary myocardial abnormality resulting in myocytic death. Similar appearances have been documented in infants with infection with coxsackievirus, and in animal models of viral myocarditis, raising the possibility that aortic valvar stenosis, and perhaps even atresia in the human fetus, are caused by the same agent.[48–51] We are currently investigating this possibility further.

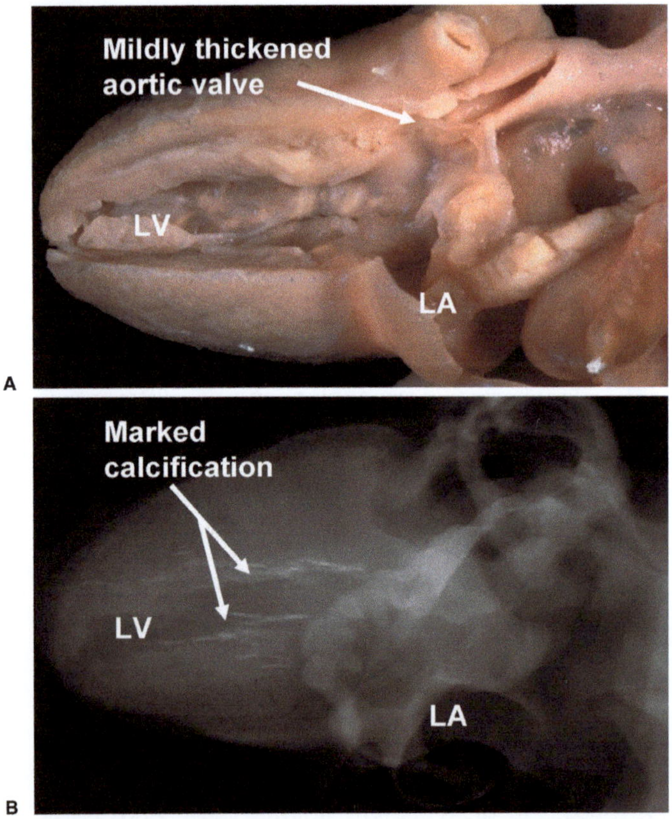

FIGURE 2.17. An example of a fetus with aortic valvular stenosis showing that left ventricular changes are often more severe than those in the aortic valve, suggesting that the primary abnormality is of the left ventricle. In this fetus, the aortic valve is only mildly thickened and there is minimal endocardial fibroelastosis (EFE) present (A), but a microfocal x-ray shows marked calcification of the left side of the ventricular septum and papillary muscles (B). LV, left ventricle; LA, left atrium.

Third, with respect to basic parameters of right ventricular size, measurements in the middle trimester fetus show that adaptation to left heart hypoplasia has already begun to occur. The right ventricles are shorter, wider, and thicker-walled than their age-related counterparts (Fig. 2.1B). There is poor correlation, nonetheless, between the amount of thickening or enlargement of the right ventricle and the degree of hypoplasia of the left heart.[35] This implies either that other mechanisms are acting to augment the output from the right heart, such as an increase in heart rate or ejection fraction, or that the right heart is still in the process of adaptation, and that output is suboptimal. Validation of the first possibility will require careful studies of right heart function during fetal life. If the latter is correct, however, then total cardiac output would be reduced, and one might expect this to manifest itself in the deleterious growth of the fetus, and in the reduced size of other downstream structures such as the great arteries. Some evidence suggesting this is indeed the case is presented below, but in terms of gross parameters of fetal size, such as weight and head size, there appears to be no significant change between fetuses with left heart hypoplasia and those with structurally normal hearts.[35] This, then, is also in keeping with anecdotal evidence that babies with hypoplasia of the left heart are normally grown at birth.[38,52,53]

THE GREAT ARTERIES AND AORTIC ARCH

The ability of the right ventricle to function adequately as the sole pumping chamber during fetal life clearly has implications for the growth and development of the entire arterial system. If it is the case that the right ventricle is, indeed, still in the process of compensating for underdevelopment of the left heart in the middle trimester of pregnancy, and the combined ventricular output is less than optimal, this should be apparent in the size of the descending aorta. When we measured the diameter of this artery in fetuses with varying forms of left heart hypoplasia, the overall z-scores were significantly reduced compared to the mean for structurally normal fetuses (Fig. 2.1A). This, then, implies that flow through the descending aorta is reduced in the mid-trimester and, perhaps, that the right ventricle has not yet fully compensated for left-sided hypoplasia, despite the changes in comformation and mural thickness already described. Surprisingly, though, the hypoplasia of the descending aorta is the most severe in fetuses with aortic valvar stenosis. The reason for this is uncertain, but one explanation is that there is compression of the right ventricle. The left ventricles in such fetuses show a variable degree of hypoplasia, but they are rounded and globular, and still of sufficient size to compress the right ventricle as a result. In consequence, the left ventricle and the ventricular septum bow toward the right ventricle, reducing its size and thereby decreasing the output from the right heart. Add to this a reduced output from the left heart, resulting from left ventricular dysfunction and aortic valvar stenosis, and total cardiac output would be decreased, leading to poor growth of the descending aorta.

The right ventricle must support the arterial supply to the head and neck, as well as to the lower body, by supplying the aortic arch and head and arm arteries in retrograde fashion via the arterial duct. Thus, arterial supply to the upper body is dependent not only on patency of the arterial duct, but also on an unobstructed pathway from the duct to the descending aorta. According to Rudolph et al.,[54] the aortic arch must be of normal size in the setting of hypoplasia of the left heart in order for the fetus to survive. Examination of fetuses suggests that, at least in the mid-trimester of pregnancy, this is not the case. First, examination of the dimensions of the aorta from its root to the aortic isthmus shows that it is significantly shorter and narrower than in the structurally normal heart, and that the hypoplasia continues from the ascending aorta, not only into the aortic arch itself, but also into the arteries supplying the head and neck (Fig. 2.1A). Secondly, almost always there is obstruction of the aortic arch between the entry of the arterial duct and the aortic arch in the form of a preductal coarctation. In fetuses with mitral and/or aortic atresia, this obstruction is frequently severe, and can involve the origin of the left subclavian artery, which often has not migrated fully across the mouth of the arterial duct (Fig. 2.2). All of these features suggest that, even though size of the fetal head seems appropriate, perfusion of the aortic arch, and therefore the head and neck, may not be optimal in the middle trimester of pregnancy. Whether this is overcome during subsequent gestation remains to be seen. In many respects this may be irrelevant, since it is in the mid-trimester of pregnancy that maturation of the cortex of the brain and migration of neurones occurs. It seems very likely that, if perfusion of the head is suboptimal, this maturation would be disrupted, leading to neurological deficit at birth. Neurological deficits have been found in neonates and infants with hypoplasia of the left heart following both transplantation and palliative surgery, but have previously been attributed to the surgical procedures.[55–57] The evidence from the fetal phenotype suggests that the deficit could instead be acquired in mid-gestation, during a window of suboptimal arterial perfusion, but this requires further investigation and correlation.

How, then, does the fetus with aortic atresia and severe preductal coarctation manage to perfuse the coronary arteries in retrograde fashion? As mentioned previously, this combination of lesions was found in virtually all fetuses with mitral and/or aortic atresia. In all, the lesion took the form of an infolding of the aortic wall, overlaid on its luminal side by a ring of muscular duct-like tissue (Fig. 2.18). In some, the lesion was less severe, allowing at least some opportunity for blood to

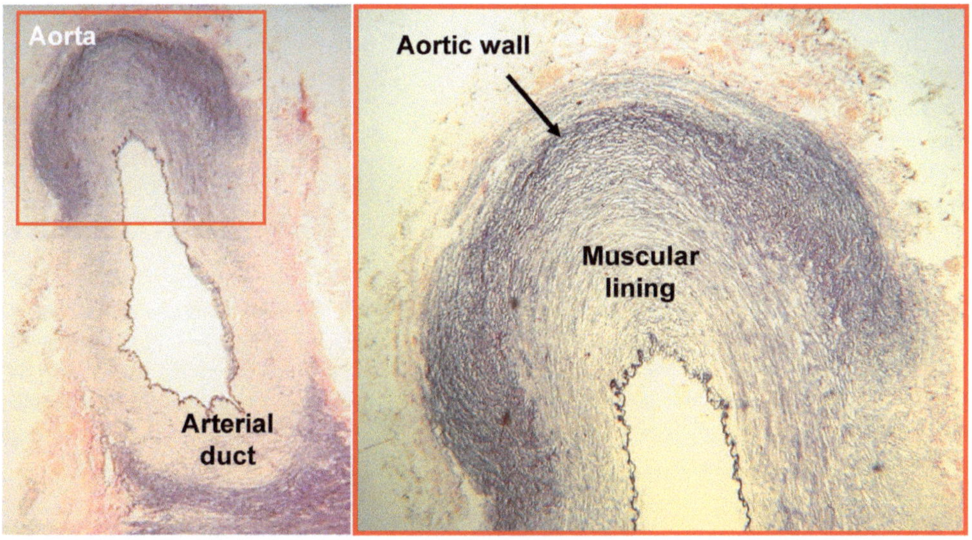

FIGURE 2.18. Sections through a coarctation lesion in a fetus with mitral and aortic atresia. In these fetuses, the coarctation ridge is composed both of infolding and thickening of the aortic wall. The thickening of the aortic wall comprises muscular tissue, which is continuous with the wall of the arterial duct.

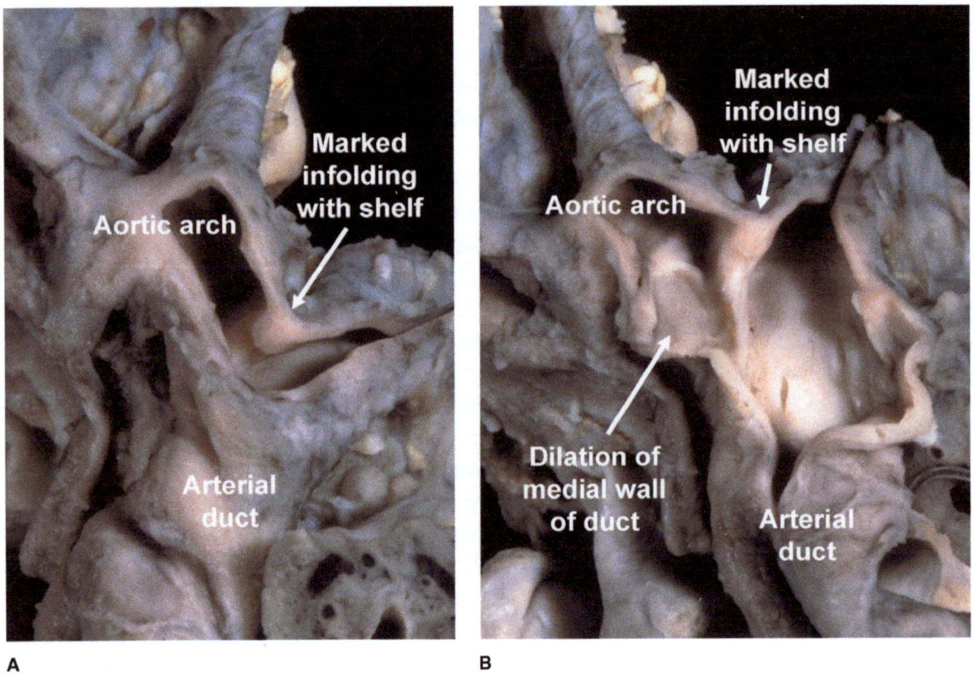

FIGURE 2.19. These two images, of the same heart with mitral and aortic atresia, show fetal adaptation to severe coarctation with the arch in closed position (A) and retracted open (B). For the fetus to survive, there must be retrograde flow from the arterial duct into the coronary arteries via the aortic arch, but this is prevented if the coarctation is severe. In fetuses that have survived to the middle trimester, the medial wall of the duct has dilated, to bypass the obstructive coarctation lesion.

cross from the arterial duct to the aortic arch. In others, however, the obstruction itself was marked. In these, the medial anterior wall of the arterial duct had become dilated so that blood could bypass the lesion (Fig. 2.19). This adaptation, then, allows for fetal survival in the presence of aortic atresia and severe preductal coarctation. These appearances are markedly different from those seen in fetuses with coarctation of the aorta in whom the aortic valve is patent. In these cases, the obstruction is mainly due to infolding of the aortic wall, with very little intimal thickening. The infolding can be so severe that there is virtually atresia of the aortic arch at the junction with the arterial duct (Fig. 2.20). Flow to the coronary arteries is still possible, providing the aortic valve and root stay patent. Given the abnormal structure of the aortic valve in up to one-half of these fetuses, and the marked hypoplasia of the aortic root present in some, particularly those with Turner's syndrome, there is probably a fine balance in coronary arterial supply in these fetuses, with concurrent risk of intrauterine death.

Of note, coarctation of the aorta was least severe in fetuses with aortic valvar stenosis. In these, the aortic arch is larger, and the coarctation shelf consisted of a shallow ridge of tissue, situated more frequently opposite the mouth of the arterial duct (Fig. 2.21). Although the ridge is muscular in nature, and therefore "duct-like," it appears to be a distinct thickening of the intima opposite the mouth of the arterial duct, sometimes sep-

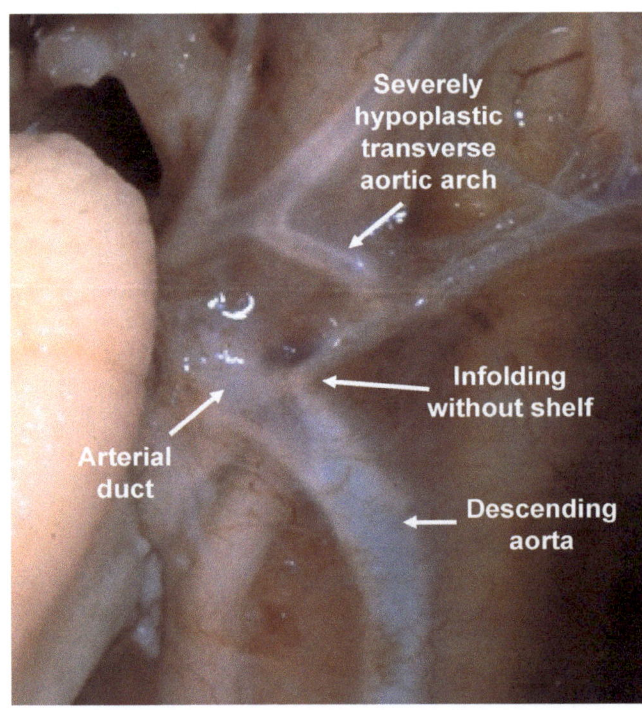

FIGURE 2.20. In contrast to the findings shown in Figures 2.18 and 2.19, the arrangement seen in fetuses with coarctation and a patent aortic root is formed mainly of an infolding in the aortic wall with little evidence of thickening. The transverse aortic arch is severely hypoplastic and has been brought forward into the mouth of the arterial duct.

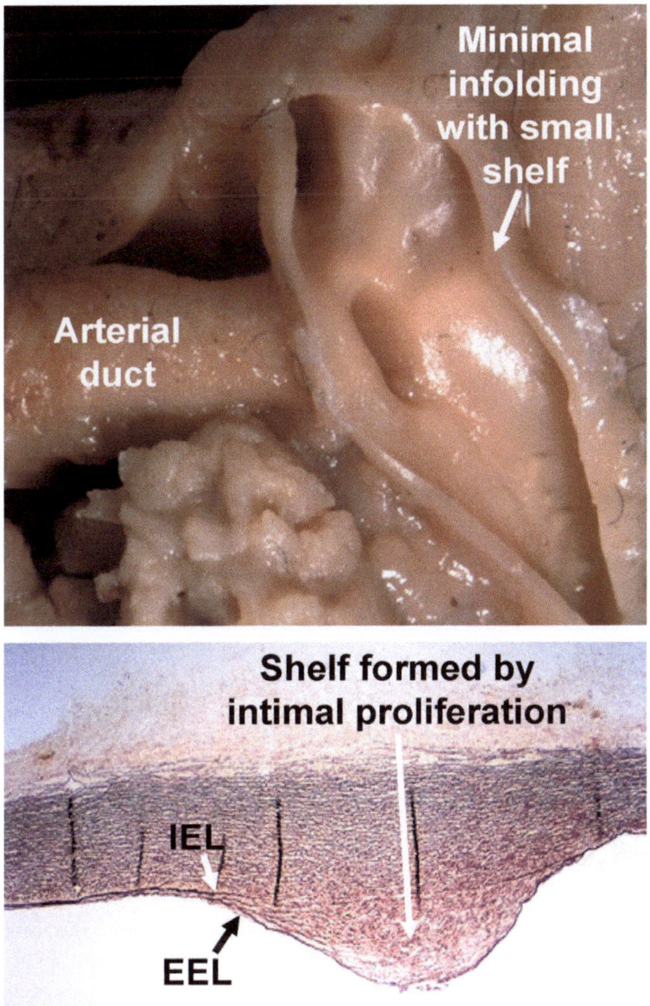

FIGURE 2.21. An example of mild coarctation of the type frequently seen in fetuses with aortic valvular stenosis. The coarctation is often opposite the mouth of the arterial duct and is shallow with little infolding of the aortic wall. On histologic sectioning it is confirmed as proliferation of muscular tissue within the intima between the internal (IEL) and external (EEL) elastic lamina. This forms a sling extending from the mouth of the arterial duct.

arated from it by normal aortic wall (Fig. 2.21). This morphology gives some clues to the formation of coarctation in left heart hypoplasia. It suggests that it is formed as a result of the aortic wall, together with an intimal shelf, being drawn into the mouth of the arterial duct as a result of atrophy and hypoplasia of the arch, rather than ductal tissue migrating out into the aortic arch. Only this mechanism can account for the appearances seen in fetuses with differing forms of left-sided hypoplasia.

OTHER ASSOCIATED ABNORMALITIES AS CLUES TO ETIOLOGY

So far, I have discussed in detail distinct parts of the phenotype of left heart hypoplasia in the human fetus. These, taken together with other associated anomalies of, for instance the pulmonary and systemic veins, lead me to the conclusion that there are several forms of hypoplasia that are likely to occur at differing stages of development, and that could well have differing etiologies. Severe forms of hypoplasia, such as mitral and aortic atresia, have a higher incidence of associated extracardiac and intracardiac anomalies (Fig. 2.22). This is to be expected, given that the cardiac anomaly itself is likely to have occurred at around 5 weeks of gestation. The finding suggests that there is a primary problem in development of the heart, and even of other organs. Less severe forms, such as aortic valvar stenosis, have very few associated cardiac and extracardiac abnormalities, but do have marked abnormalities of the left ventricular myocardium that are not explainable on the basis of the aortic valvar lesion. This suggests that this form of hypoplasia could be acquired later in fetal life, following septation and formation of the heart, possibly as a result of a left ventricular cardiomyopathy of the type seen later in life subsequent to infection with the coxsackie B3 virus.[51] In the remaining groups, it is difficult, currently, to suggest a unique cause. Fetuses with aortic valvar atresia clearly overlap in part with those with aortic valvar stenosis, and also with those having mitral atresia due to an imperforate mitral valve in conjunction with aortic atresia. The mechanism could equally involve abnormal expansion of the left atrioventricular or ventriculoarterial junctions, as it could be due to the presence of left ventricular dysfunction. Similarly, in fetuses with primary abnormalities of the aortic arch, there is clearly abnormal development of the aortic arch, but left ventricular hypoplasia is a more variable feature. Its presence may also be due to abnormalities in expansion of the left atrioventricular junction or atrial septum. Overall, there is little correlation between the size of the atrial septum and the dimensions of the left heart. It well may be that, as we look earlier in development, we will find that delayed perforation of the primary septum is a common pathway that leads to many forms of left-sided hypoplasia, but that subsequently this primary abnormality is masked by changes in haemodynamics. Only future examination of very early human fetuses, or the development of animal models of the disease, is likely to answer these questions.

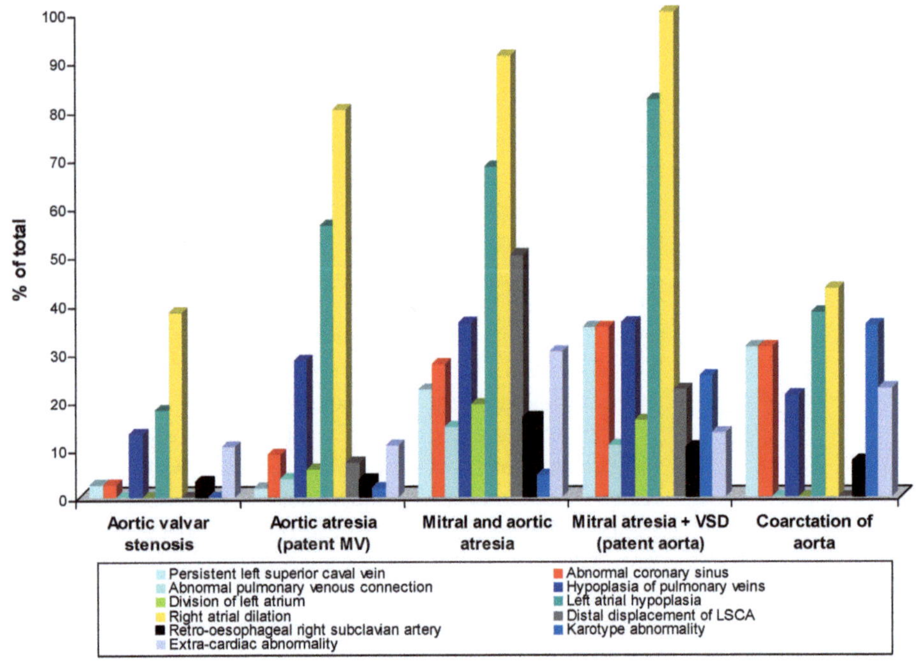

Figure 2.22.

References

1. Manhoff LJ. Congenital heart disease: tricuspid atresia and mitral atresia associated with transposition of the great vessels. Am Heart J 1945;29:90.
2. Shaner RF. Malformation of the atrioventricular cushions of the embryo pig and its relations to defects of the conus and truncus arteriosus. Am J Anat 1949; 84:431–455.
3. Shaner RF. Anomalies of the heart bulbus. J Pediatr 1962;61:233–241.
4. Watson DG, Rowe RD, Coren PE, Duckworth JWA. Mitral atresia with normal aortic valve. Report of 11 cases and review of the literature. Pediatrics 1960;25:450–467.
5. Neill CA, Tuerk J. Aortic atresia, hypoplasia of the ascending aorta and underdevelopment of the left ventricle. In: Watson H, ed. Paediatric Cardiology. London: Lloyd-Luke, 1968.
6. Sherman D. An Atlas of Congenital Heart Disease. Philadelphia: Lea and Febiger, 1963.
7. Lumb G, Dawkins WA. Congenital atresia of mitral and aortic valves with vestigial left ventricle (three cases). Am Heart J 1960;60:378–387.
8. Donnally HH. Congenital mitral stenosis: report of a case of developmental mitral stenosis combined with hypoplasia of left ventricle and left auricle; rudimentary aorta and other developmental defects. JAMA 1924;82:1318.
9. Bharati S, Lev M. The surgical anatomy of hypoplasia of aortic tract complex. J Thorac Cardiovasc Surg 1984; 88:97–101.
10. Remmell-Dow DR, Bharati S, Davis JT, Lev M, Allen HD. Hypoplasia of the eustachian valve and abnormal orientation of the limbus of the foramen ovale in hypoplastic left heart syndrome. Am Heart J 1995;130:148–152.
11. Kreysig FL. Die Krankheiten des Herzens Systrematisch Bearbeitet und Durch eigne Beobachtungen Earlautert. Berlin: 1817.
12. Rauchfuss C. Die angeborenen Entwicklungsfehler des Herzens und die Ftalkrankheiten des Herzens und der Grossen Gefasse. In: Gerhardt's Handbuch der Kinderkrankheiten. Tubingen; H. Lauppschen Buchhandlung, 1878.
13. Farber S, Hubbard J. Fetal endomyocarditis: intrauterine infection as the cause of congenital cardiac anomalies. Am J Med Sci 1933;186:705–713.
14. Gross P. Concept of fetal endocarditis: a general review with report of an illustrative case. Arch Pathol 1941;31:163–177.
15. Fishman NH, Hof RB, Rudolph AM, Heymann MA. Models of congenital heart disease in fetal lambs. Circulation 1978;58:354–364.
16. Harh JY, Paul MH, Gallen WJ, Friedberg DZ, Kaplan S. Experimental production of hypoplastic left heart syndrome in the chick embryo. Am J Cardiol 1973;31:51–56.
17. Sedmera D, Hu N, Weiss KM, Keller BB, Denslow S, Thompson RP. Cellular changes in experimental left heart hypoplasia. Anat Rec 2002;267:137–145.
18. Allan LD, Anderson RH, and Cook AC. Atresia or absence of the left-sided atrioventricular connection in the fetus: echocardiographic diagnosis and outcome. Ultrasound Obstet Gynecol 1996;8:295–302.
19. Silverman NH, Enderlein M, Golbus M. Ultrasonic recognition of aortic valve atresia in utero. Am J Cardiol. 1984;53:391–392.

20. Royston P, Wright EM. How to construct "normal ranges" for fetal variables. Ultrasound Obstet Gynecol 1998;11:30–38.
21. Altman DG, Chitty LS. Design and analysis of studies to derive charts of fetal size. Ultrasound Obstet Gynecol 1993;3:378–384.
22. Altman, DG, and Chitty, LS. Charts of fetal size. 1. Methodology. J Obstet Gynaecol 1994;101:29-34.
23. Royston P, Altman DG. Regression using factional polynomials of continuous covariates: parsimonious parametric modelling. Appl Statist 1994;43:429-467.
24. Allan LD, Sharland G, Tynan MJ. The natural history of the hypoplastic left heart syndrome. Int J Cardiol 1989;25:341–343.
25. Danford DA, Cronican P. Hypoplastic left heart syndrome: progression of left ventricular dilation and dysfunction to left ventricular hypoplasia in utero. Am Heart J 1992;123:1712–1713.
26. Simpson JM, Sharland GK. Natural history and outcome of aortic stenosis diagnosed prenatally. Heart 1997; 77:205–210.
27. Allan LD, Sharland GK, Milburn A, et al. The prospective diagnosis of 1006 consecutive cases of congenital heart disease in the fetus. J Am Coll Cardiol 1994;23:1452–1458.
28. Huggon IC, Cook AC, Smeeton NC, Magee AG, Sharland GK. Atrioventricular septal defects diagnosed in fetal life: associated cardiac and extra-cardiac abnormalities and outcome. J Am Coll Cardiol 2000;36:593–601.
29. Rose V, Clarke E. Etiology of congenital heart disease. In: Freedom RM, Benson LN, Smallhorn JF, eds. Neonatal Heart Disease. London: Springer-Verlag, 1992:5.
30. Hornberger LK. Abnormalities of the atrioventricular junction: b. Mitral valve anomalies. In: Allan LD, Hornberger LK, Sharland GK, eds. Textbook of Fetal Cardiology. London: Greenwich Medical Media, 2000:148–162.
31. Rudolph AM. The changes in the circulation after birth: their importance in congenital heart disease. Circulation 1970;41:343–359.
32. Gittenberger-De-Groot AC, Wenink ACG. Mitral atresia: morphological details. Br Heart J 1984;51:252–258.
33. Rigby ML, Anderson RH, Gibson D, Jones ODH, Joseph MC, Shinebourne EA. Two-dimensional echocardiographic categorisation of the univentricular heart. Ventricular morphology, type, and mode of atrioventricular connection. Br Heart J 1981;46:603–612.
34. Thiene G, Daliento L, Frescura C, De Tommasi M, Macartney FJ, Anderson RH. Atresia of left atrioventricular orifice. Anatomical investigation in 62 cases. Br Heart J 1981;45:393–401.
35. Cook AC. The anatomy of left heart hypoplasia in the human fetus. Ph.D. thesis, University of London, 2003.
36. Aiello VD, Ho SY, Anderson RH, Thiene G. Morphologic features of the hypoplastic left heart syndrome-a reappraisal. Pediatr Pathol 1990;10:931–943.
37. Mickell JJ, Mathews RA, Anderson RH, et al. The anatomical heterogeneity of hearts lacking a patent communication between the left atrium and the ventricular mass ("mitral atresia") in presence of a patent aortic valve. Eur Heart J 1983;4:477–486.
38. Fyler DC, Buckley LP, Hellenbrand WE, Cohn HE. Report of the New England Regional Infant Cardiac Program. Pediatrics 1980;65 (suppl):375–461.
39. Boon AR, Roberts DF. A family study of coarctation of the aorta. J Med Genet 1976;13:420–433.
40. Tchervenkov CI, Tahta SA, Jutras LC, Bjorkland MJ. Biventricular repair in neonatal hypoplastic left heart complex. Ann Thorac Surg 1998;66:1350–1357.
41. O'Connor WN, Cash JB, Cottrill CM, Johnson GL, Noonan JA. Ventriculocoronary connections in hypoplastic left hearts: an autopsy microscopic study. Circulation 1982;66:1078–1086.
42. Sauer U, Gittenberger De Groot AC, Geishauser M, Babic R, Buhlmeyer K. Coronary arteries in the hypoplastic left heart syndrome. Histopathologic and histometrical studies and implications for surgery. Circulation 1989;80:I168–I176.
43. Freedom RM, Culham JAC, Moes CAF, Harrington DP. Selective aortic root angiography in the hypoplastic left heart syndrome. Eur J Cardiol. 1976;4:25–29.
44. Sandor GS, Cook AC, Sharland GK, Ho SY, Potts JE. Coronary arterial abnormalities in pulmonary atresia with intact ventricular septum diagnosed during fetal life. Cardiol Young 2002;12:436–444.
45. Kasznica J, Ursell PC, Blanc WA, Gersony WM. Abnormalities of the coronary circulation in pulmonary atresia and intact ventricular septum. Am Heart J 1987;114:1415–1420.
46. McKay R, Smith A, Leung MP, Arnold R, Anderson RH. Morphology of the ventriculoaortic junction in critical aortic stenosis. Implications for hemodynamic function and clinical management. J Thorac Cardiovasc Surg 1992;104:434–442.
47. Lakier JB, Lewis AB, Heymann MA, Stanger P, Hoffman JI, Rudolph AM. Isolated aortic stenosis in the neonate. Natural history and hemodynamic considerations. Circulation 1974;50:801–808.
48. Reyes MP, Ho KL, Smith F, Lerner AM. A mouse model of dilated-type cardiomyopathy due to coxsackievirus B3. J Infect Dis 1981;144(3):232–236.
49. Reyes MP, Lerner AM. Coxscakie myocarditis—with special reference to acute and chronic effects. Prog Cardiovasc Dis 1985;27:373–394.
50. Barson WJ, Craenen J, Hosier DM, Brawley RL, Hilty MD. Survival following myocarditis and myocardial calcification associated with infection by Coxsackie Virus B-4. Pediatrics 1981;68:79–81.
51. Goren A, Kaplan M, Glaser J, Isacsohn M. Chronic neonatal Coxsackie myocarditis. Arch Dis Child 1989;64:404–406.

52. Benson L, Freedom RM, Gersony W, Gundry SR, Sauer U, Boucek M. Session II: cardiac replacement in infants and children: indication and limitations. J Heart Lung Transplant 1991;10:791–801.
53. Freedom RM, Benson LN, Smallhorn JF. Hypoplastic Left Heart Syndrome. In: Freedom RM, Benson LN, Smallhorn JF, eds. Neonatal Heart Disease. London: Springer-Verlag, 1992:339.
54. Rudolph AM, Heymann MA, Spitznas U. Hemodynamic considerations in the development of narrowing of the aorta. Am J Cardiol 1972;30:514–525.
55. Glauser TA, Rorke LB, Weinberg PM, Clancy RR. Congenital brain anomalies associated with the hypoplastic left heart syndrome. Pediatrics 1990; 85:984–990.
56. Olson DM, Shewmon DA. Electroencephalographic abnormalities in infants with hypoplastic left heart syndrome. Pediatr Neurol 1989;5:93–98.
57. Rogers BT, Msall ME, Buck GM, et al. Neurodevelopmental outcome of infants with hypoplastic left heart syndrome. J Pediatr 1995;126:496–498.

Chapter 3

Antenatal Diagnosis of Hypoplastic Left Heart Syndrome

Gurleen Sharland

Hypoplastic left heart syndrome encompasses a spectrum of lesions affecting the left heart, the main features of which are hypoplasia of the left ventricle, together with atresia or stenosis of its inlet and outlet. The positional relationship of the great arterial trunks in these cases is normal. Such a syndrome, involving underdevelopment of the left atrium, left ventricle and ascending aorta, was already recognised in the 1800s.[1–4] These early investigators noted that the right side of the heart and the pulmonary trunk were disproportionately large, although the heart itself was of normal size, and that the cavity of the left ventricle was unnaturally small. In 1952, Lev[5] used the term "hypoplasia of the aortic tract complex" to describe a group of abnormalities with inflow and outflow obstruction of the left heart. Then, in 1958, the term "hypoplastic left heart syndrome" was proposed by Noonan and Nadas[6] to include obstructive lesions on the left side of the heart associated with a hypoplastic left ventricle and right ventricular hypertrophy. In 1968, Sinha et al.[7] argued that the constellation would better be called the hypoplastic left ventricle syndrome, thus excluding all cases of primary coarctation and interrupted aortic arch. Despite the wisdom of this suggestion, the term "hypoplastic left heart syndrome" has persisted, albeit that there is still controversy as to precisely which lesions should be included.

Most commonly, it is accepted that the syndrome consists of aortic atresia with or without mitral atresia, although some cases will have mitral atresia with a patent aortic valve and the aorta arising concordantly from the left ventricle. In a minority of cases, the mitral and aortic valves may both be patent, but the left heart structures are very hypoplastic. Those placed in this latter group fall into the severe end of the spectrum of coarctation of the aorta as seen during fetal life.

The incidence of the syndrome is reported to vary from 4.8% to 9%, and the prevalence is between 0.1 and 0.27 cases in every 1000 live births.[8–14] Babies born with this condition appear normal and healthy at birth, and develop signs of heart failure as the arterial duct starts to close. When this happens, the systemic circulation is inadequate, and the baby develops circulatory collapse, presenting in a shock-like state. Early diagnosis of this condition thus has potential benefits, and antenatal diagnosis can be made as early as 14 weeks of pregnancy.

FETAL SPECTRUM AND DIAGNOSIS

Hypoplastic left heart syndrome is one of the commonest forms of cardiac abnormality encountered in fetal life.[15,16] It accounts for almost one-fifth of structural cardiac abnormalities in the large series of fetuses scanned at Guy's Hospital (Fig. 3.1). The reason for this high incidence is that fetuses with the syndrome are readily detectable by examination of the four-chamber view of the heart during routine obstetric ultrasonic examination.[17] In the majority of cases referred to our unit, such an abnormality in the four chamber view had been noted at the time of the routine obstetric scan. In a smaller proportion of cases, there was a different reason for referral (Fig. 3.2). Although this type of cardiac abnormality can be detected from 13 to 14 weeks of pregnancy, the majority of cases are detected at between 20 and 22 weeks, coinciding with the timing of the obstetric scan seeking congenital anomalies (Fig. 3.3). Some cases, however, are not detected and referred until the third trimester of pregnancy (Fig. 3.3). Unfortunately, there are still some cases that remain undetected until after birth.[18]

Key to abbreviations for all figures: AA, ascending aorta; AL, aortic leaflet of the mitral valve; CS, coronary sinus; D, arterial duct; E, endocardial fibroelastosis; IA, innominate (brachiocephalic) artery; ICV, inferior caval vein; IS, isthmus; LA, left atrium; LAA, left atrial appendage; LCC, left common carotid artery; LPA, left pulmonary artery; LPV, left pulmonary vein; LS, left subclavian artery; LV, left ventricle; MV, mitral valve; OF, oval fossa; PT, pulmonary trunk; PV, pulmonary valve; RA, right atrium; RAA, right atrial appendage; RPA, right pulmonary artery; RPV, right pulmonary vein; RV, right ventricle; S, septomarginal trabeculation; SCV, superior caval vein; TV, tricuspid valve.

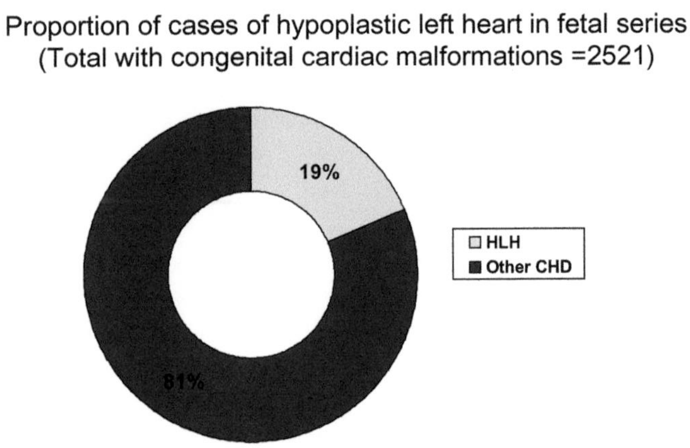

FIGURE 3.1. The proportion of cases with hypoplastic left heart (HLH) syndrome in the series of 2521 fetuses with congenital heart disease (CHD) seen in the fetal cardiology unit at Guy's Hospital.

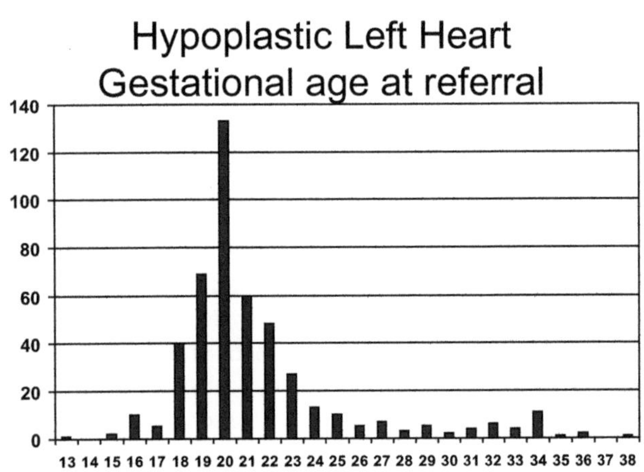

FIGURE 3.3. The gestational age at referral in 471 fetuses with hypoplastic left heart syndrome.

The fetal echocardiographic findings are those of an abnormal left ventricle and aorta. The left ventricle is small, although there is some variation in this depending on the patency of the mitral valve, and whether there is an associated ventricular septal defect. In cases of aortic and mitral atresia, the left ventricle is tiny.

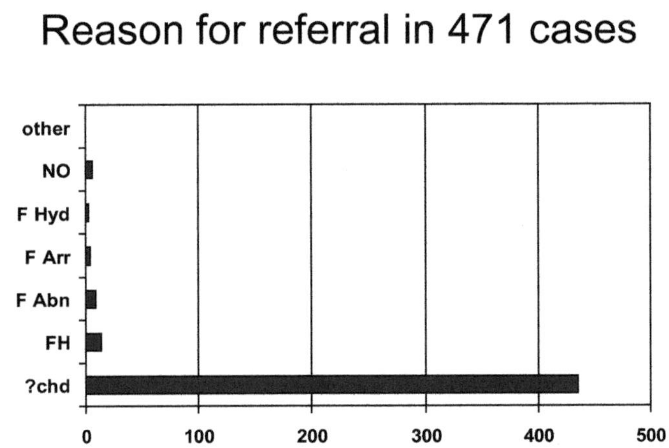

FIGURE 3.2. Reasons for referral in 471 cases of hypoplastic left heart syndrome diagnosed during fetal life. ?chd, abnormality suspected during routine screening; FH, family history of congenital heart disease; F Abn, extracardiac fetal abnormality; F Arr, fetal arrhythmia; F Hyd, non immune fetal hydrops; NO, increased nuchal translucency.

Indeed, oftentimes it is not discernible (Fig. 3.4). In this setting, the components of the left heart are so underdeveloped that, in the four-chamber view, the impression may be of only one atrium, one atrioventricular valve, and one ventricle. In cases of aortic atresia associated with a patent mitral valve and an intact interventricular septum, the left ventricular cavity is more easily recognised, but it is often echogenic, globular, and dysfunctional (Fig. 3.5A). In these cases the normal inflow of the left ventricle cannot be demonstrated, even though the mitral valve is patent (Fig. 3.5B). The echogenicity of the left ventricle correlates well with the finding of endocardial fibroelastosis at postmortem examination of the heart. The prerequisite for the presence of endocardial fibroelastosis is severe obstruction to the egress of blood from the left ventricle in the presence of an intact ventricular septum, but with a route for ingress of mitral flow.

If there is aortic atresia, then the ascending aorta is usually tiny, measuring around 1 mm in the middle-trimester fetus. This makes it difficult to demonstrate, but careful searching usually allows identification of the hypoplastic vessel compared to the larger pulmonary trunk (Fig. 3.6). No forward flow is detectable across the aortic valve, and only retrograde flow from the duct is detected in the hypoplastic aortic arch (Fig. 3.7). In some cases, the aortic valve may be patent, as in the setting of mitral atresia with a ventricular septal defect and the aorta arising concordantly from the left ventricle. In these instances, the aortic root may be normal in size, or only slightly hypoplastic. Forward flow will then usually be detected in the ascending aorta. A minority of cases fall in the severe end of the spectrum of coarcta-

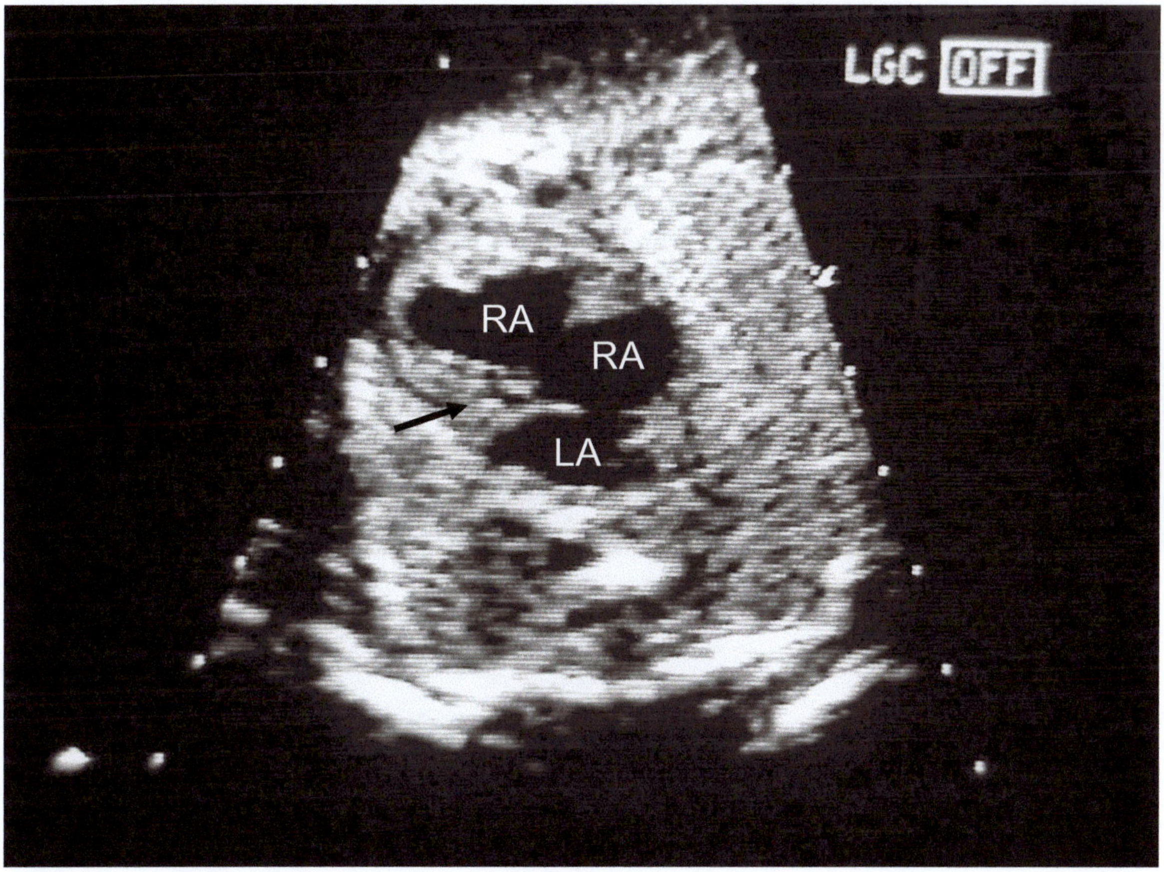

Figure 3.4. The four-chamber view of the fetal heart in a baby with mitral and aortic atresia. The left ventricle (arrow) is slit like and hypoplastic. LA, left atrium; RA, right atrium; RV, right ventricle.

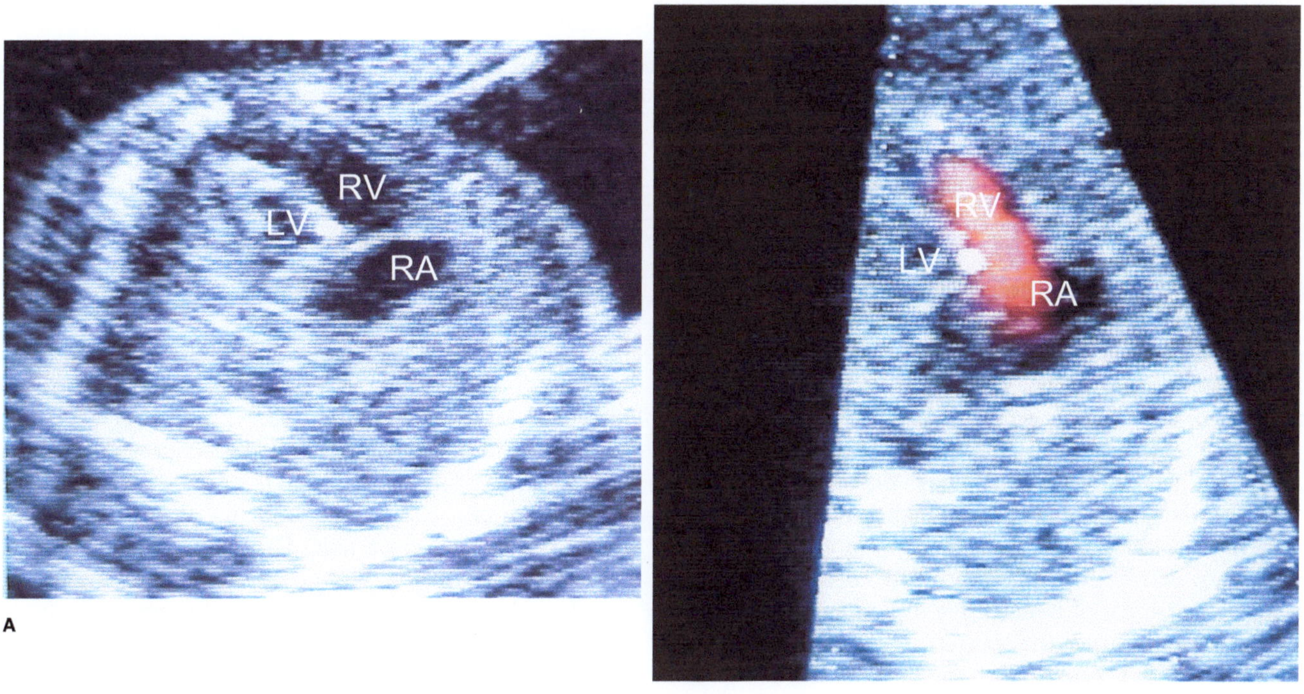

Figure 3.5. The four-chamber appearance in aortic atresia with a patent but hypoplastic mitral valve. The left ventricle appears echogenic and globular (A). It was poorly contractile on the moving image. Colour flow demonstrates flow across the tricuspid valve into the right ventricle, but no flow is seen across the mitral valve into the left ventricle (B). LV, left ventricle; RA, right atrium; RV, right ventricle.

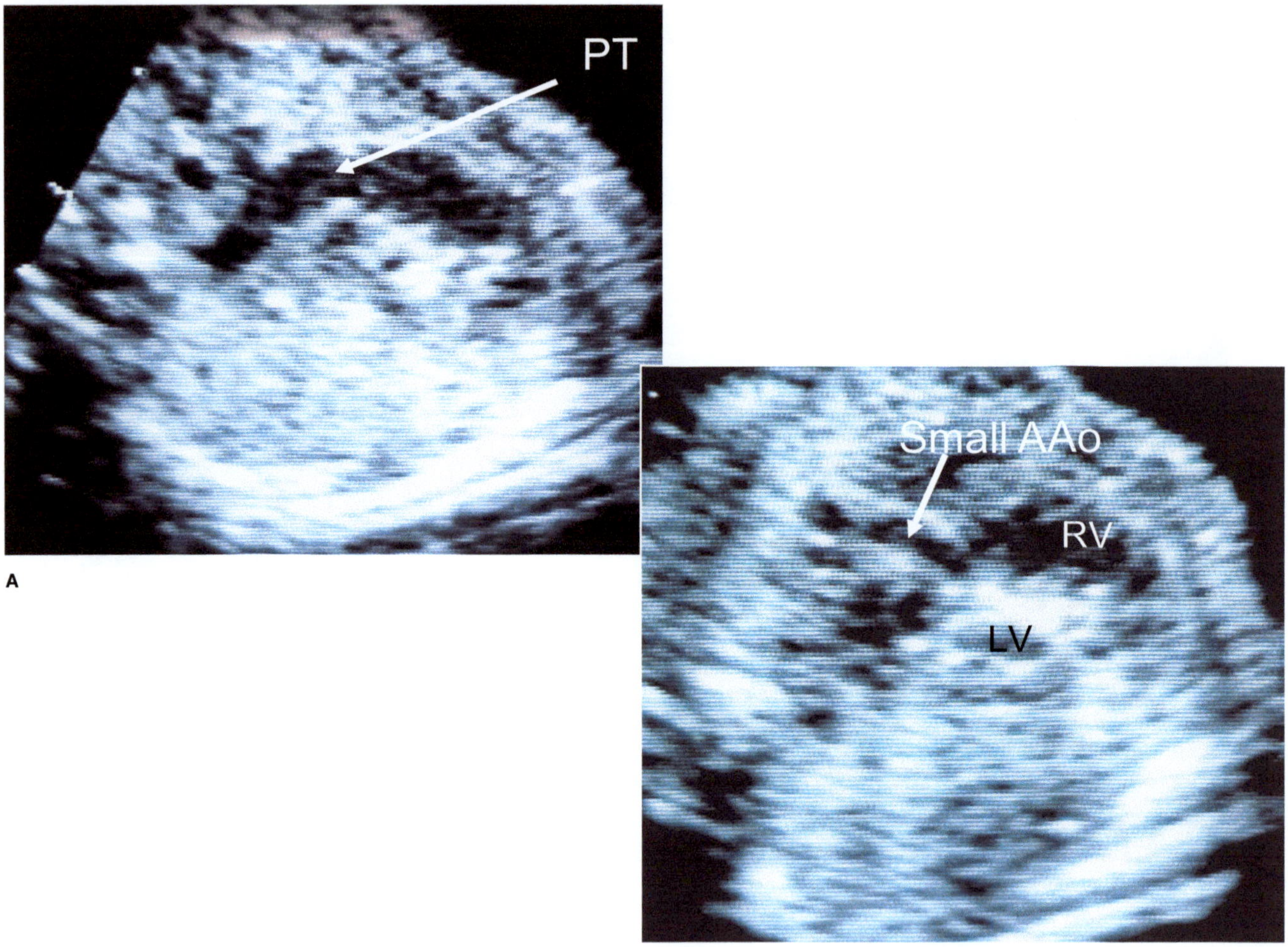

FIGURE 3.6. The pulmonary trunk (A) is significantly larger than the ascending aorta (B) in this example. The aorta is seen arising from a fibrotic and hypoplastic left ventricle in B. AAo, ascending aorta; PT, pulmonary trunk; LV, left ventricle; RV, right ventricle.

tion of the aorta, where there is marked hypoplasia of all the components of the left heart, but the mitral and aortic valves are still patent (Fig. 3.8).

Premature closure of the oval foramen has been suggested as a cause of the syndrome, but in the majority of fetuses diagnosed with this lesion, the foramen is usually patent.[19] In these instances, the blood usually shunts from left to right atriums, which is the reverse of the normal situation (Fig. 3.9). In a minority of cases, the foramen can be restrictive, or more rarely intact. A clue to the presence of a restrictive atrial septum can be the finding of an abnormal pulsed Doppler pattern of pulmonary venous flow.[20] An increase in the reverse flow wave into the pulmonary veins during atrial systole correlates with restriction of the foramen.

ASSOCIATED SYNDROMES

Those with the syndrome can occasionally have chromosomal anomalies, particularly Turner's syndrome, but also trisomy 18 and 13.[21–23] It has also been reported to be associated with duplication of the short arm of chromosome 12.[24] Extracardiac structural malformations can also occur, and in some instances this increases the surgical risk. Brackley et al.[25] reported karyotypic anomalies in one-eighth of their cases, and associated extracardiac anomalies in one-fifth. In the fetuses we have seen at Guy's Hospital, from 3% to 4% had an associated karyotypic anomaly, and a further 7% had an extracardiac anomaly in association with a normal karyotype.

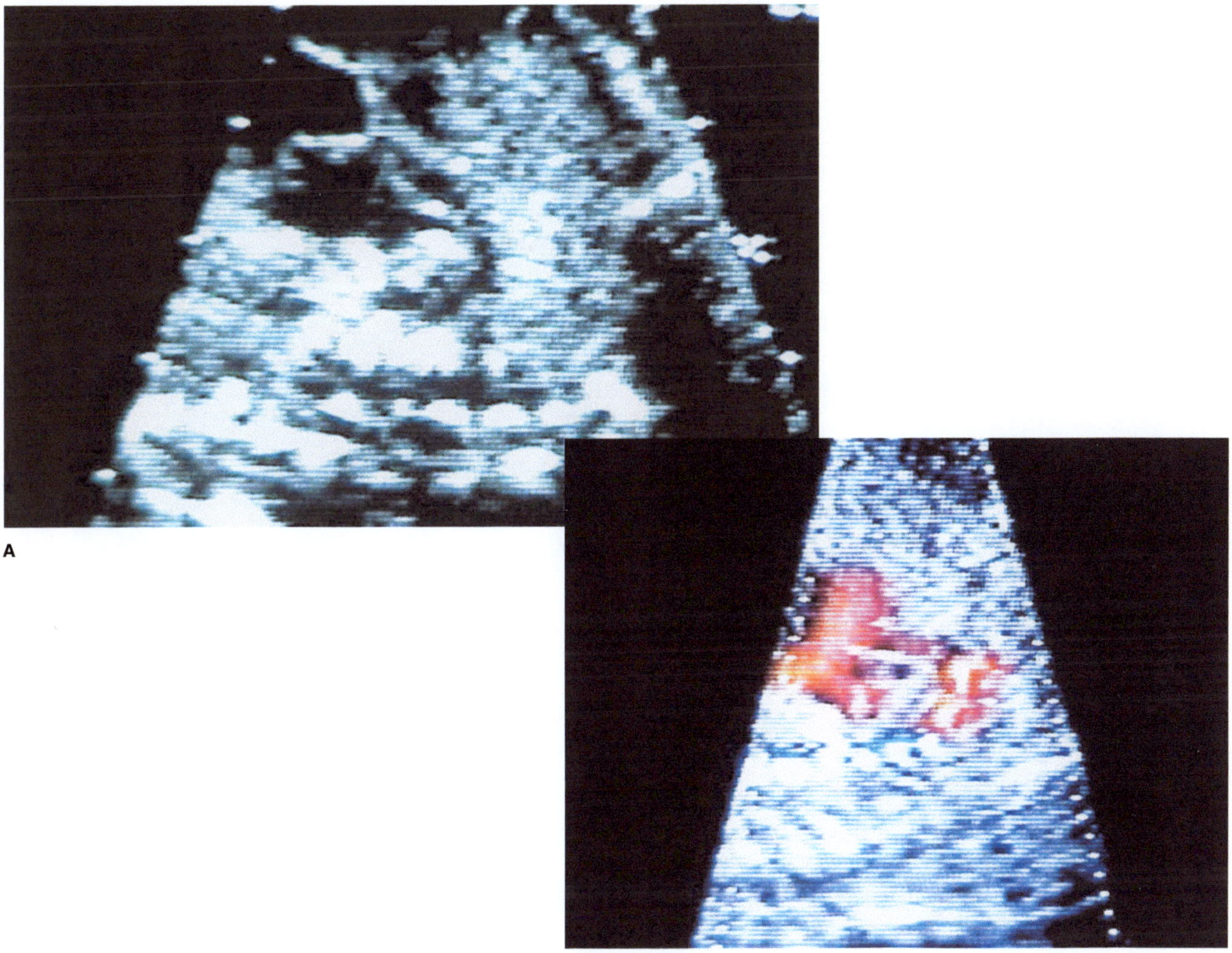

A

B

FIGURE 3.7. A very hypoplastic aortic arch (A), with reversed flow only in the arch, shown in red (B).

MANAGEMENT

Hypoplastic left heart syndrome lies at the severe end of the spectrum of cardiac malformations. When the diagnosis is made in the middle trimester of pregnancy or earlier, the parents have the option of either interrupting the pregnancy or proceeding to term. The therapeutic options that are available after birth include no treatment, cardiac transplantation, or palliative surgery. The option of no treatment, allowing nature to take its course, is a complex decision made between physicians and parents. If transplantation is being considered, then the major limiting factor is the availability of a donor heart.[26] Such an organ may become available within hours, but may take weeks or months. Most of the time, it may not become available at all. Palliative surgery involves progression through the stages of the Norwood procedure,[27,28] which is described in detail in other chapters of this book. The Norwood strategy has now become the preferred treatment option for this condition in the United Kingdom. In the United States, survival from the first stage now exceeds 80% in some centres,[29] with even better results reported elsewhere in this book (see Chapter 7). Current results of this strategy in the United Kingdom are now beginning to approach those achieved in the United States. Some studies have reported that the outcome for babies with a prenatal diagnosis is worse, with the mortality following the first stage of the Norwood procedure being in the range of 50% to 55%, with some babies not even reaching the operating room.[30] Other studies, in contrast, have reported an improved outcome in babies with a prenatal diagnosis,[31] as well as finding a reduction in early neurological morbidity compared with cases

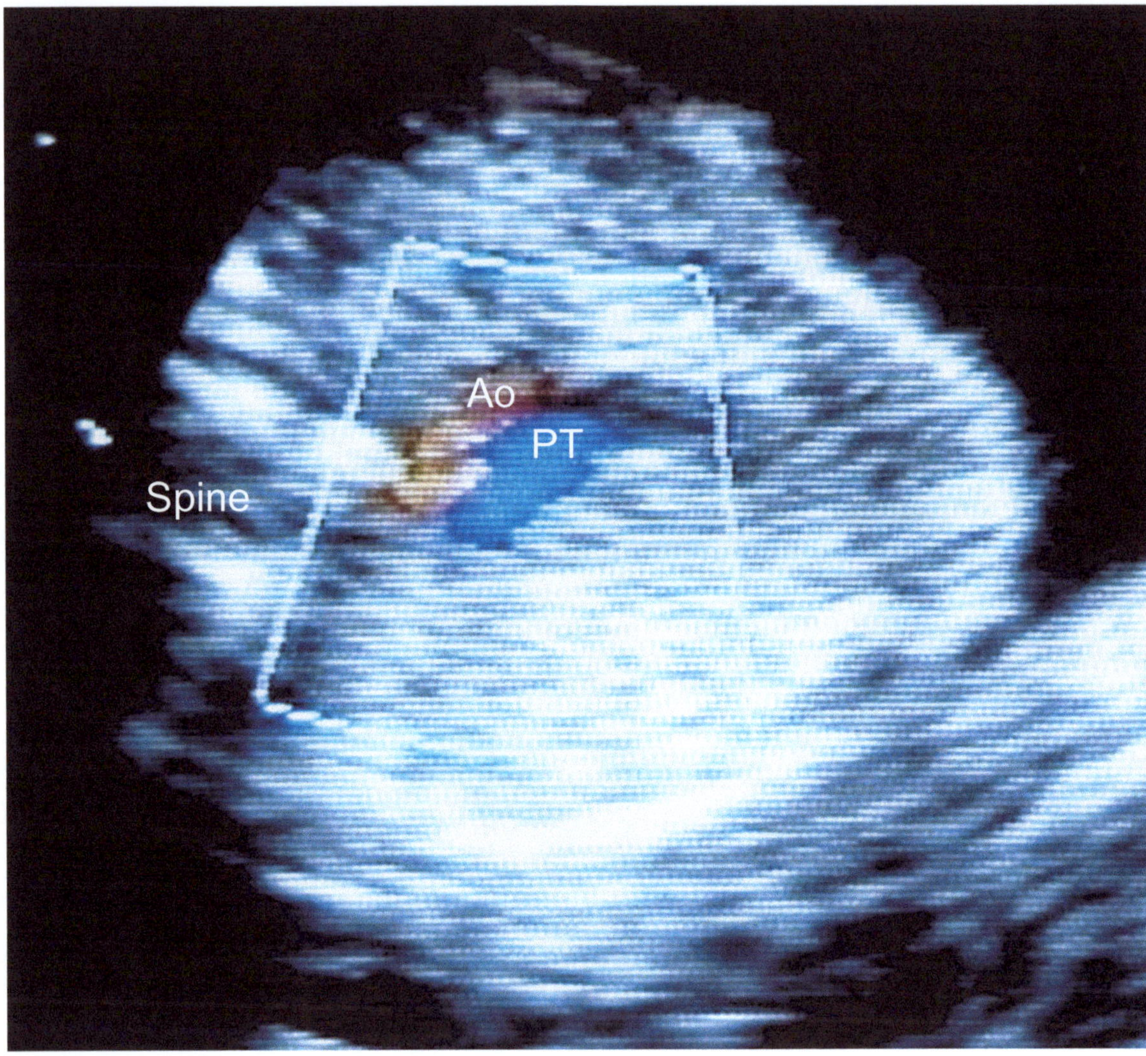

C

FIGURE 3.7. (*cont.*) In a transverse view of the pulmonary trunk and aortic arch (C), there is forward flow in the pulmonary trunk (PT), seen in blue heading toward the spine, and retrograde flow in the aortic arch (Ao), seen in red heading away from the spine.

presenting postnatally.[32] If antenatal data are analysed on an "intention to treat" basis, the results are different compared with looking at surgical mortality alone.[30,33] This is because some pregnancies may result in a spontaneous intrauterine loss, and some infants may die postnatally prior to surgery. These factors have to be taken into account, in addition to the surgical mortality, when counselling parents antenatally.

It is now apparent that, with a properly organised programme, it is possible to achieve satisfactory results in the early and medium term. Uncertainty remains, however, as to what will happen to the systemic right ventricular function over the long term. Whether a significant proportion of the survivors of the staged Norwood approach will eventually require cardiac transplantation also remains unknown. Thus, when discussing the options following prenatal diagnosis, a realistic view must be given about the outcome, not just in terms of surgical mortality, but also in terms of longer-term development and survival. As yet, there is little information about the long-term survival for these children. Despite the uncertainty of the long-term outlook for the Norwood procedure, many parents are now electing to choose this option. Since the introduction of the Norwood procedure in the United Kingdom, there has been a change in the proportion of pregnancies resulting in termination. Overall, however, the majority of parents still choose the option of termination of pregnancy following prenatal diagnosis. The overall outcome of pregnancy following prenatal diagnosis in

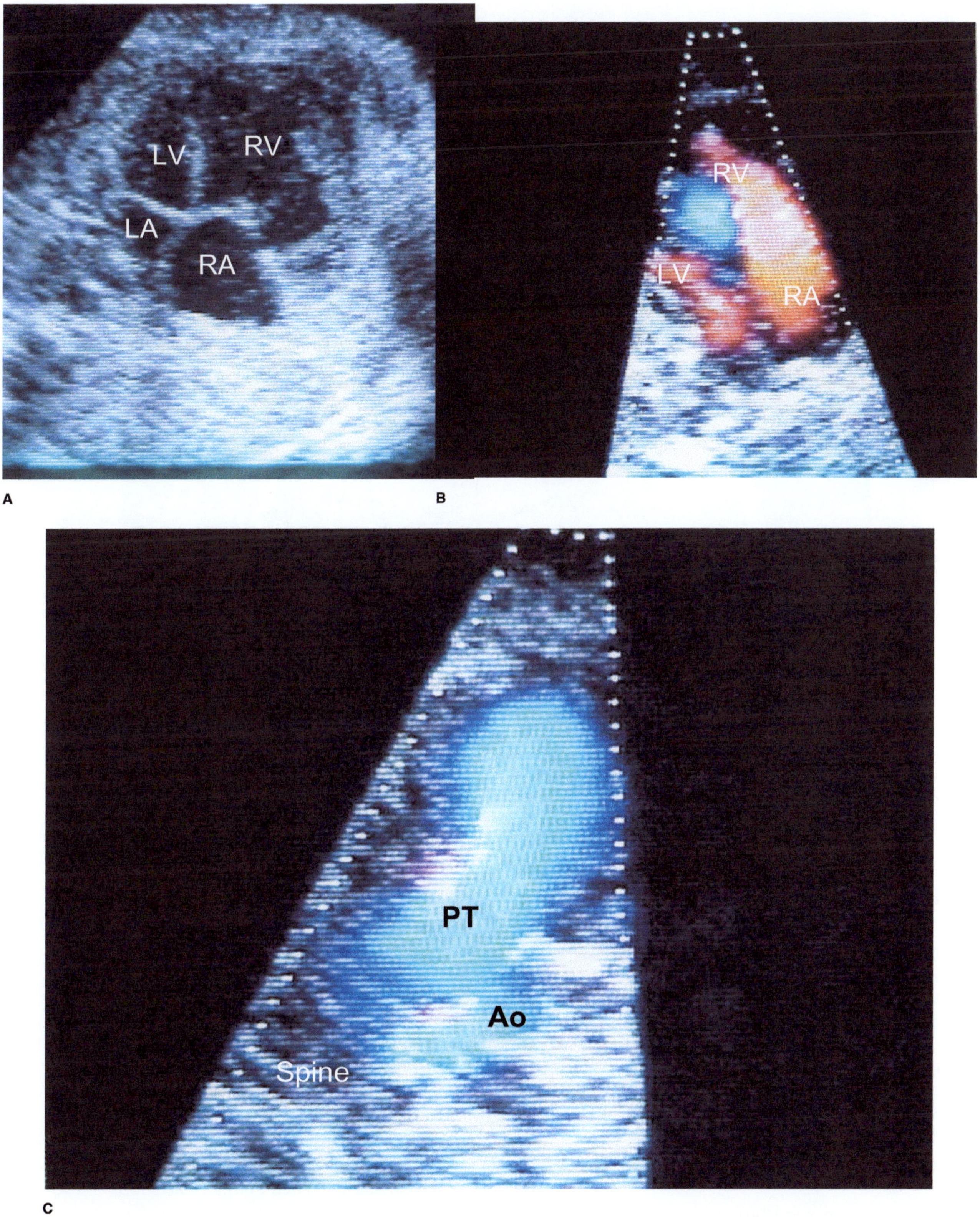

FIGURE 3.8. An example of a four-chamber view in severe coarctation falling into the pattern of hypoplastic left heart syndrome. The left atrium (LA) and left ventricle (LV) are significantly smaller than the right atrium (RA) and right ventricle, respectively (RV) (A). Colour flow mapping reveals that the mitral valve is patent (B). A view of the great arteries in the same baby is shown in C. Although the aorta (Ao) is significantly smaller than the pulmonary trunk (PT), there is still forward flow in the arch.

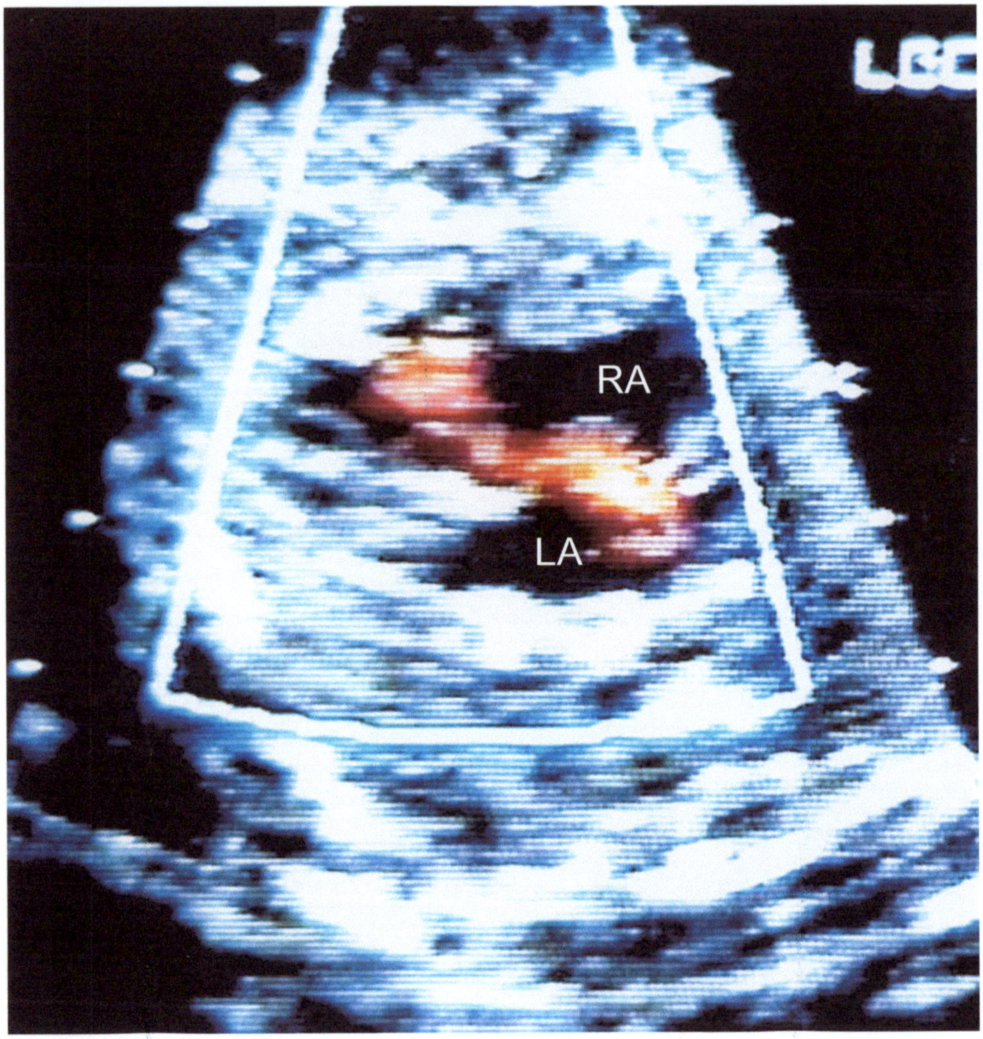

FIGURE 3.9. The interatrial flow in this example is left to right, which is the reversal of the normal pattern. LA, left atrium; RA, right atrium.

the fetuses with the syndrome seen at Guy's Hospital is shown in Figure 3.10. It is worth noting that the percentage of parents electing to choose termination has fallen from 78% in 1994, prior to the introduction of the Norwood strategy at our unit, to 63% in 2001. In comparison, the percentage of parents choosing to interrupt the pregnancy following prenatal diagnosis of congenital cardiac disease in general is currently between 35% and 40%.

CONCLUSION

Hypoplastic left heart syndrome can be diagnosed with a high degree of accuracy from 13 to 14 weeks of pregnancy. Antenatal diagnosis gives parents time to understand, and come to terms with, the problem in the heart of their baby. It also gives them the opportunity to make informed choices, and to be prepared for events postnatally.

References

1. Canton M. Congenital obliteration of origin of the aorta. Trans Pathol Soc Lond 1850;2:38.
2. Shattock SG. Atresia of the aortic aperture in an infant. Trans Pathol Soc Lond 1891;32:38–39.
3. Dilg J. Ein Beitrag zur Kenntniss seltener, Herzanomalian im Ancschluss an einen Fall von angeborner linksseitiger Conusstenose. Arch Pathol Anat 1883;91:193–259.
4. Martens G. Zwei Falle von Aortenatresie. Arch Pathol Anat 1890;121:322–335.
5. Lev M. Pathological anatomy and interrelationship of the hypoplasia of the aortic tract complexes. Lab Invest 1952;1:61–70.

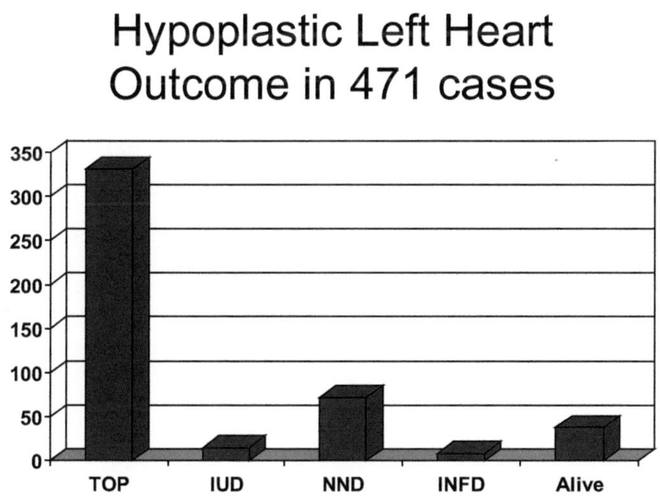

FIGURE 3.10. The outcome in 471 cases of hypoplastic left heart syndrome diagnosed during fetal life. TOP, termination of pregnancy; IUD, intrauterine death; NND, neonatal death; INFD, infant death.

6. Noonan JA, Nadas AS. The hypoplastic left heart syndrome. An analysis of 101 cases. Pediatr Clin North Am 1958;5:1029–1056.
7. Sinha SN, Rusnak SL, Sommers HM, Cole RB, Muster AJ, Paul MH. Hypoplastic left ventricle syndrome. Analysis of thirty autopsy cases in infants with surgical considerations. Am J Cardiol 1968;21:166–73.
8. Fyler DC, Buckley LP, Hellenbrand WE, Cohn HE. Report of the New England Regional Infant Cardiac Care Program. Pediatrics 1980;65(suppl):375–461.
9. Izukawa, T, Mulholland HC, Rowe RD, et al. Structural heart disease in the newborn. Changing profile: comparison of 1975 and 1965. Arch Dis Child 1979;54:281–285.
10. Scott DJ, Rigby ML, Miller GAH, Shinebourne EA. The presentation of symptomatic heart disease in infancy based on 10 years experience, 1973–1982. Br Heart J 1984;52:248–257.
11. Ferencz C, Rubin JD, McCarter RJ, et al. Congenital heart disease: prevalence at livebirth. Am J Epidemiol 1985;121:31–36.
12. Freedom RM. Atresia or hypoplasia of the left atrioventricular and/or ventriculo-arterial. In: Anderson RH, McCartney FJ, Shinebourne EA, Tynan M, eds. Paediatric Cardiology. Edinburgh, Melbourne, London, New York, Churchill, Livingstone. 737–765.
13. Calgren LE. The incidence of congenital heart disease in children born in Gothenburg 1941–1950. Br Heart J 1959;21:40–50.
14. Mitchell SC, Korones SB, Berendes HW. Congenital heart disease in 56,109 births. Incidence and natural history. Circulation 1971;43:323–332.
15. Allan LD, Sharland GK, Milburn A, et al. Prospective diagnosis of 1,006 consecutive cases of congenital heart disease in the fetus. J Am Coll Cardiol 1994;23:1452–1458.
16. Allan LD, Sharland G, Cook A. Hypoplastic left and right ventricles. In: Color Atlas of Fetal Cardiology. London: Mosby-Wolfe, 1994;83–93.
17. Sharland GK, Allan LD. Screening for congenital heart disease prenatally: results of a 3-year study in the South East Thames Region. Br J Obstet Gynaecol 1992;99:220–225.
18. Bull C, on behalf of British Paediatric Cardiac Association. Current and potential impact of fetal diagnosis on prevalence and spectrum of serious congenital heart disease at term in the UK. Lancet 1999;345: 1242–1247.
19. Cook AC. The anatomy of left heart hypoplasia in the human fetus. Ph.D. thesis, University of London, 2003.
20. Better DJ, Apfel HD, Zidere V, Allan LD. Pulmonary venous flow patterns in the hypoplastic left heart syndrome. Heart 1999;81:646–650.
21. Ferencz C, Neill CA, Boughman JA, Rubin JD, Brenner JI, Perry LW. Congenital cardiovascular abnormalities associated with chromosome abnormalities: an epidemiologic study. J Pediatr 1989;114:79–86.
22. Wladimiroff JW, Stewart PA, Reuss A, Sachs ES. Cardiac and extracardiac anomalies as indicators for trisomies 13 and 18: a prenatal ultrasound study. Prenatal Diagn 1989;9:515–520.
23. Van Egmond H, Orye E, Praet M, Coppens M, Devloo-Blancquaert A. Hypoplastic left heart syndrome and 45X karyotype. Br Heart J 1988;60:69–71.
24. Oazi QH, Kanchanapoomi R, Cooper R, Madahar C, Beller E. Brief clinical report. Dup (12p) and hypoplastic left heart. Am J Med Genet 1981;9:195–199.
25. Brackley KJ, Kilby MD, Wright JG, et al. Outcome after prenatal diagnosis of hypoplastic left heart syndrome: a case series. Lancet 2000;356:1143–1147.
26. Stuart AG, Wren C, Sharples PM, Hunter S, Hey EN. Hypoplastic left heart syndrome: more potential transplant recipients than suitable donors. Lancet 1991;337:957–959.
27. Norwood WI, Kirklin JK, Snaders SP. Hypoplastic left heart syndrome: experience with palliative surgery. Am J Cardiol 1980;45:87–91.
28. Norwood WI, Lang P, Hansen DD. Physiologic repair of aortic atresia–hypoplastic left heart syndrome. N Engl J Med 1983;308:23–26.
29. O'Kelly SW, Bove EL. Hypoplastic left heart syndrome: terminal care is not the only option. Br Med J 1997;314:87–88.
30. Allan LD, Apfel HD, Printz BF. Outcome after prenatal diagnosis of the hypoplastic left heart syndrome. Heart 1998;79:371–374.
31. Tworetzsky W, McElhinney DB, Reddy VM, Brook MM, Hanley FL, Silverman NH. Improved surgical outcome after fetal diagnosis of hypoplastic left heart syndrome. Circulation 2001;103:1269–1273.

32. Mahle WT, Clancy RR, McGaurn SP, Goin JE, Clark BJ. Impact of prenatal diagnosis on survival and early neurologic morbidity in neonates with the hypoplastic left heart syndrome. Pediatrics 2001;107:1277–1282.
33. Andrews R, Tulloh R, Sharland G, et al. Outcome of staged reconstructive surgery for hypoplastic left heart syndrome following antenatal diagnosis. Arch Dis Child 2001;85:474–477.

Chapter 4

THE GENETICS OF HYPOPLASIA OF THE LEFT HEART

Helen Cox and David I. Wilson

Congenital heart disease affects 8 babies in every thousand born alive, and is one of the major causes of infant mortality and morbidity. At the time of diagnosis, the major issues for parents relate to survival and outcome. Sooner or later, however, they will have concerns regarding the aetiology of the cardiac defect. Establishing the aetiology of congenital malformations becomes extremely important when parents ask:

- "What caused it?"
- "Why did it happen to us?"
- "Will it happen again, either to us or to our children?"

These questions can be answered accurately only if the aetiologies are understood and clearly identified. The answers become especially important for "lethal" malformations, such as hypoplasia of the left heart, that carry high rates of morbidity and mortality. If there is potential for a genetic aetiology, then a significant risk of recurrence may influence future reproductive decisions of the parents.

It seems likely that hypoplasia of the left heart is aetiologically heterogeneous, with both environmental and heritable factors contributing to the phenotype. This chapter discusses the evidence that a proportion of individuals with hypoplasia of the left heart have a genetic aetiology, and the impact that this might have on them and their families.

EXISTING GENETIC EVIDENCE

There is an increase in the risk of recurrence of congenital heart disease in families where one individual has hypoplasia of the left heart. For a couple who have had one child with a nonsyndromic congenital cardiac malformation, the probability that it will happen again in a subsequent pregnancy, called sibling recurrence risk, is approximately 2% to 3%.[1] For parents of a child with hypoplasia of the left heart, the sibling risk of recurrence is significantly above the "population risk" of one in 5000 live births, and has been estimated to be as high as 13.5%,[2] although no large prospective studies have been reported.

Patterns of Inheritance of Hypoplasia of the Left Heart

Patterns of inheritance have been difficult to determine because of the severity of the phenotype, but available evidence suggests that there may be more than one pattern. Familial recurrence has been reported in consanguineous marriages with affected pairs of siblings, and this suggests an autosomal-recessive pattern.[3] Autosomal-dominant inheritance has been suggested for nonsyndromic obstructive left-sided lesions,[4] and there are reports of an increased incidence of valvular lesions in parents of children with the hypoplasia of the left heart. Brenner and colleagues[5] found that one-eighth of children had one parent with a bicuspid aortic valve and/or aortic stenosis. This is consistent with an autosomal-dominant pattern of inheritance with a variable phenotype, as has been seen in other conditions such as the DiGeorge and Holt-Oram syndromes, both of which include cardiac malformations. Transmission of hypoplasia of the left heart from parents to offspring, however, has not been reported, although this may not be surprising for such a severe phenotype that was lethal prior to the advent of cardiopulmonary bypass

TABLE 4.1. Chromosomal abnormalities associated with hypoplasia of the left heart (HLH)

Chromosomal abnormality and birth incidence	*% CHD in newborns*	*% HLH*
45,X (Turner syndrome) 1/5000	23%[12]	15%[12]
47,XX/XY+13 (Patau syndrome) 1/12,000	82%	
47,XX/XY+18 (Edwards syndrome) 1/5000	85%	~6%
47,XX/XY+21 (Down syndrome) 1/650	40%	
46,XX/XYdel(22)(q11) (DiGeorge syndrome) 1/4000	~60%	Low
46,XX/XYdel(11)q	65%	13%

CHD, congenital heart disease.

surgery. The observation of Brenner and colleagues,[5] nonetheless, raises the possibility that an autosomal-dominant gene exists that is associated with a much wider phenotype.

Specific Chromosomal Abnormalities are Associated with Hypoplasia of the Left Heart

Chromosomal analysis of individuals with hypoplasia of the left heart usually proves normal, although several associated chromosomal anomalies have been specifically reported (Table 4.1). Of these the most notable are Turner's syndrome, or 45,X, and the common trisomies of chromosomes 13, 18, and 21.

As recorded in the human cytogenetics database,[6] one-eighth of all 11q23 deletions have hypoplasia of the left heart, and an additional one-tenth have mitral stenosis, which may represent the mild end of a left-sided obstructive phenotype. Computational analyses of the material contained within this database, calculating the association of single congenital malformations with cytogenetic regions, revealed that 11q23 has a very significant association with hypoplasia of the left heart.[7,8] This was confirmed independently,[8] and more recently a critical region was defined within 11q25 by investigating overlapping deletions of different sizes in 3 children with hypoplasia of the left heart.[9] This approach has narrowed the gene or genes causing the cardiac defect to within the terminal 8 megabases of 11q25.

Hypoplasia of the Left Heart Occur as Part of Genetically Determined Syndromes

Hypoplasia of the left heart has been reported as part of several other syndromes (Table 4.2), and they may have important implications for prognosis. In this context, it is important to define the term "syndrome" as a recognisable pattern of malformations or disease that typically affects multiple systems of organs or tissues. Although hypoplasia of the left heart is frequently referred to as "hypoplastic left heart syndrome," in the strictest sense it represents maldevelopment of only one organ, namely the heart, and therefore should be grouped as a nonsyndromic cardiac malformation.

Although most of the syndromes listed in Table 4.2 are clinically distinct and easily recognisable, this may not be the case if hypoplasia of the left heart is diagnosed prenatally. The aetiological basis for some of these syndromes has yet to be determined. For others, specific genes have already been implicated, such as *TBX5* with Holt-Oram syndrome, and *CRB* (CRE binding protein gene) with Rubinstein-Taybi syndrome. This is further proof that the phenotypes making up hypoplasia of the left heart can result from mutations in more than 1 gene.

Animal Models of Human Disease

The advent of transgenic technology in animals, particularly in mice, in which targeted genes may be either switched off or overexpressed, has provided significant insight into the molecular mechanisms of congenital malformations. Over the past 10 years, many genes have been disrupted. The resulting murine phenotypes include those with cardiac defects resembling or identical to those in humans. This is not surprising, given the similarities in mammalian embryonic development. The challenge will be to use these clues to determine the genes in which dysfunction or mutation causes cardiac malformations in humans.

PRACTICAL CLINICAL IMPLICATIONS

Whilst the investigation of the genetic basis of hypoplasia of the left heart is intellectually challenging, what are the implications for the parents and family of a child? The need for evidence to answer the many questions arising at the time of diagnosis is great. In addition, as rates of survival improve for children treated surgically, attention to factors influencing the quality of life increases.

TABLE 4.2. Syndromes and the Online Mendelian Inheritance in Man (OMIM) number for each, in which hypoplasia of the left heart has been described (number of reported cases, where known); main clinical features, mode of inheritance, locus to which the syndrome maps, and the causative gene are tabulated

Holt-Oram syndrome *142900*	Symmetrical radial ray abnormalities	Natowicz et al. 1988[13]	Dominant	12q21 *TBX5*
Smith-Lemli-Opitz syndrome (4) *270400*	Genital abnormalities, learning disability	Natowicz et al. 1988[13]	Recessive	7q32.1
Ellis–van Creveld syndrome *225500*	Postaxial polydactyly, skeletal abnormalities	Schinzel 1983[6]	Recessive	4p16
Saldino-Noonan syndrome *263560*	Short ribs, polydactyly, skeletal abnormalities	Johnson et al. 1982[14]	Recessive	
Apert syndrome (1) *101200*	Craniosynostosis, syndactyly	Natowicz et al. 1988[13]	Dominant	10q25 *FGFR2*
Cerebrocostomandibular syndrome (1) *117650*	Severe micrognathia and posterior "rib gap"	Kirk and Ades 1998[15]	Recessive/dominant	
Beckwith syndrome *130650*	Overgrowth, macroglossia, abdominal wall defects	Greenwood et al. 1975[16]	Dominant	11p15
CHARGE syndrome *214800*	Coloboma, choanal atresia, ear abnormalities, developmental delay, genital anomalies	Hall 1979,[17] Cryan et al. 1987[18]		
FG syndrome *305450*	X-linked mental retardation, macrocephaly, dysmorphism		X-linked recessive	Xq12-21?
Day-Salvatore-McLean (1)	Skeletal abnormalities, blepharophimosis	Day-Salvatore and McLean 1998[19]		
Gianotti (2 sibs) *600460*	Cleft palate, genital anomalies, ectrodactyly	Gianotti et al. 1995[20]	Recessive	
Holzgreve *236110*	Renal agenesis, cleft palate, skeletal anomalies	Holzgreve 1984[21]		
Hurst	Leptomeningeal angiomatosis, absent olfactory tracts, clefts	Hurst et al. 1992[22]		
Kaufman (2 sibs)	Vertebral and renal anomalies	Kaufman et al. 1972[23]	Recessive	
Kennerknecht *202660*	Agonadism, multiple internal malformations	Kennerknecht et al. 1995[24]	Recessive	
McPherson-Clemens (2 sibs) *601165*	Cleft lip and palate, intestinal malrotation	Mc Pherson and Clemens 1996[25]	Recessive	
Medeira	Neural tube defect, cleft lip and palate, limb reduction	Medeira et al. 1994[26]		
Neish-Roberts(2 sibs)	Nephromegaly, distinctive facies	Neish and Roberts 1992[27]		
Short rib polydactyly syndrome type 3	Small thorax, polydactyly		Recessive	
SHORT syndrome *269880*	Rieger anomaly, short stature, characteristic facies		Dominant, recessive	
X-linked cardiac valvular dysplasia *314400*	Congenital cardiac valve dysplasia	Newbury-Ecob et al. 1993[28]	X-linked recessive	
Rubinstein Taybi syndrome (3) *268600*	Distinctive facies, broad halluces and thumbs	Bartsch et al. 1999,[29] Hannauer et al. 2002[30]	Dominant	16p13 *CRB*
Toriello-Carey *217980*	Agenesis corpus callosum, Robin sequence, facial anomalies, hypotonia	Czarnecki et al. 1996[31]	? X-linked	

Prenatal Diagnosis—Investigating the Cause and Prognosis of the Affected Fetus

Hypoplasia of the left heart, in some centres, can now be diagnosed prenatally in over nine-tenths of affected fetuses.[10] Genetic advice regarding management of both the pregnancy and the parents is most important in this situation. Most centres offer fetal karyotypic analysis, along with detailed sonographic investigation to determine whether there are additional congenital malformations.[11]

Fetal karyotypic analysis, usually obtained by sampling the amniotic fluid, may provide valuable information regarding the cause of the problem, and thus the prognosis and risk of recurrence in future pregnancies. In the series of Brackley et al.,[11] one-twentieth of cases that had karyotypic analysis had aneuploidy as the cause of the left-sided hypoplasia. If aneuploidy is the cause, the couple may be offered a low risk of recurrence, in addition to details about the general noncardiac prognosis.

Detailed fetal ultrasonic examination may also provide vital information. Apart from obvious implications for management, the presence and nature of additional extracardiac malformations might suggest a syndromic diagnosis, or arouse suspicion of a specific chromosomal abnormality prior to receiving the result of karyotypic analysis.

If no chromosomal abnormality is found, and sonography shows other structural malformations, how might this affect fetal prognosis, and long-term outcome? Brackley et al.[11] studied a series of cases diagnosed prenatally, and reported that the frequency of additional congenital anomalies was increased above that in the general population, although insufficient longitudinal data were available to provide prognostic advice.

If a chromosomal abnormality is not visible, and the sonographic scan is normal, what are the chances that a syndromic diagnosis will be made postnatally, with implications for the health of the child? In our own detailed clinical study, we assessed 44 infants and children with normal chromosomes and no additional anomalies. None of these had a recognised syndromic diagnosis, although one patient had a dysmorphic facial appearance suggestive of a an undiagnosed syndrome. These findings are consistent with most cases of hypoplasia of the left heart being nonsyndromic. Thus, although published reports associate the cardiac phenotype with many syndromes (Table 4.2), most are limited to single cases.

Postnatal Diagnosis—Investigating the Cause and Prognosis of the Afflicted Infant

When a neonate presents with hypoplasia of the left heart, questions and investigations relating to aetiology often take less precedence over decisions regarding surgical or nonsurgical management. This may be unfortunate on some occasions, as there may be a short window of time in which to obtain samples for investigation. Consultation with clinical geneticists may be advised if the child has multiple congenital anomalies in case specific diagnostic investigations are required, as in the case of Smith-Lemli-Opitz syndrome.

Chromosomal analysis may also be indicated, or more specific molecular cytogenetic analysis. As in the case of prenatal diagnosis, the results of these tests could answer questions relating to aetiology, prognosis, and the risk of recurrence. Although reported, left heart hypoplasia is an uncommon association with the DiGeorge or velocardiofacial syndrome phenotypes. Additional investigation for 22q11 deletion by fluorescent chromosome in situ hybridisation, therefore, is unlikely to show a deletion. Nevertheless, should additional features suggesting 22q11 deletion be observed, such as cleft palate, velopharyngeal insufficiency, or evidence of maldevelopment of the thymus or parathyroid glands, further analysis may be offered, as the diagnosis carries important implications in management for both the child and the family, such as the likelihood of deafness or a renal anomaly, and an increased risk of recurrence.

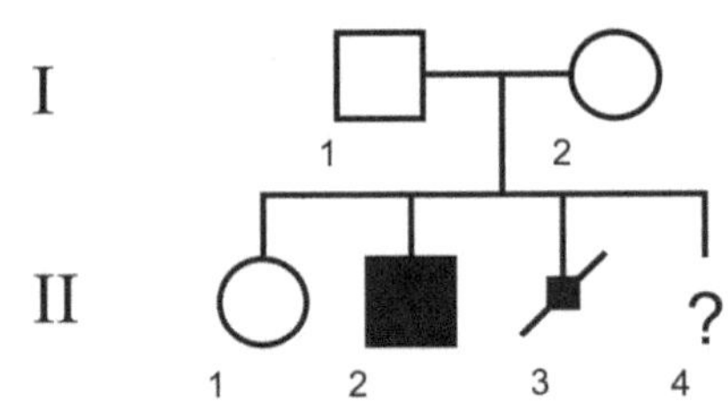

FIGURE 4.1. A healthy, nonconsanguineous couple with no family history of congenital heart disease wishes to know the chances of a future child (II.4) having hypoplasia of the left heart. Both of their affected boys have had isolated hypoplasia of the left heart, and have normal chromosomes. Patient II.2 is surviving, but III.3 was a male fetus that did not survive. This pedigree could be consistent with autosomal-recessive inheritance, with a risk of recurrence of 25%; X-linked recessive inheritance, with a risk of 25%; submicroscopic chromosomal rearrangement, a cryptic translocation for example, or autosomal-dominant inheritance with a variable phenotype. It is also consistent with a nongenetic cause for the phenotype. For a couple that have had 2 children with hypoplasia of the left heart, the empirical risk for the recurrence of any type of congenital cardiac malformation is likely to be between 15% and 20%. This, however, is an averaged "group risk" such that the risk for an an individual couple may be as high as 25%, for autosomal-recessive, or 50%, for autosomal-dominant inheritance. Detailed antenatal echocardiography was offered.

Unbalanced chromosomal rearrangements, resulting in the loss or gain of specific chromosomal regions, may be the cause of the phenotype. Indeed, the identification of a critical region within 11q25 would suggest that translocations should be specifically investigated, as they can be associated with a significant risk of recurrence depending on the precise chromosomal rearrangement.

As with prenatally diagnosed patients, there is an increase in incidence of extracardiac malformations. In our unpublished postnatal series, 6 of 50 children with normal chromosomes had additional congenital anomalies, with 5 of the 6 requiring surgery. One patient was thought to have a possible syndrome, with potential adverse developmental consequences. For any fetus or child with hypoplasia of the left heart who does not survive, postmortem examination may identify unsuspected congenital malformations that could have implications for counselling the parents regarding the aetiology and the risk of recurrence.

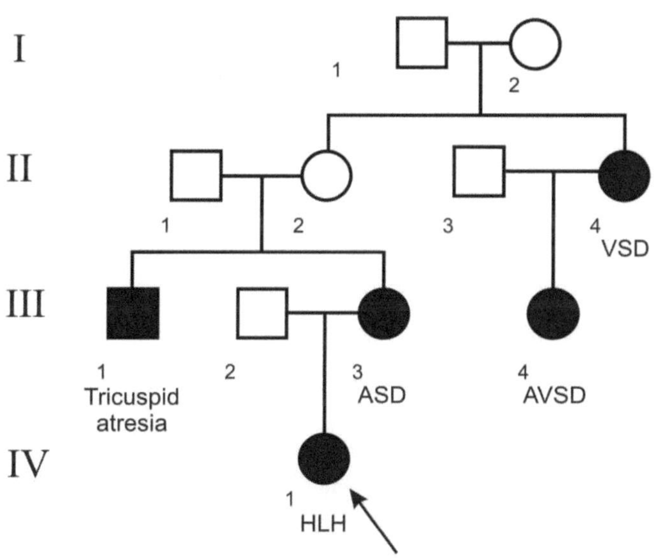

FIGURE 4.2. Our patient, II.2, wishes to know the cause of congenital heart disease in her family. Her children are nondysmorphic and have normal karyotypes. The family history suggests that a dominant gene with variable penetrance is the cause of the congenital cardiac anomalies in this family. Our patient should be offered echocardiography to exclude asymptomatic congenital cardiac disease. Patient III.1 also wishes to know the chances of having a child with a congenital cardiac anomaly. The chance of him transmitting the unidentified gene causing the cardiac phenotype in his family is 50%. The pedigree illustrates that gene carriers can have a phenotype ranging from an apparently normal heart to hypoplasia of the left heart. Although this complicates risk assessment, at a practical level detailed fetal echocardiography should be offered in all at-risk pregnancies.

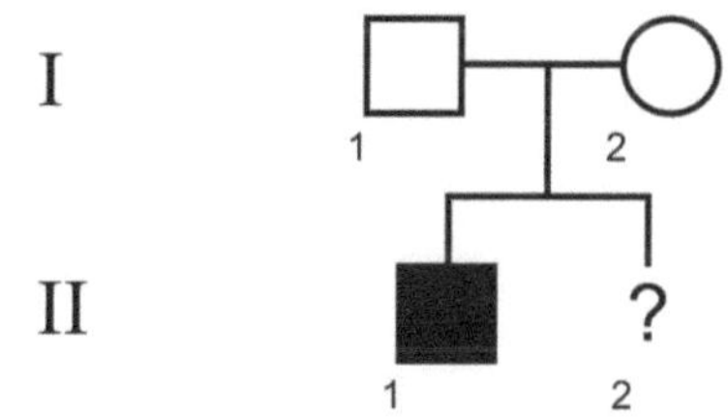

FIGURE 4.3. A healthy, nonconsanguineous couple wishes to know the chances of recurrence of hypoplasia of the left heart. There is no family history of congenital heart disease. Their son has isolated hypoplasia of the left heart and normal chromosomes, with no 22q11 deletion. It is estimated that recurrence of congenital cardiac anomalies occurs in about one-tenth of pregnancies where one previous child has hypoplasia of the left heart, albeit that this is not based on large prospective studies. It is unclear whether the cause of cardiac defect is genetic or nongenetic. There are reports of dominant inheritance of left-sided obstructive lesions with variable penetrance.[4] It is possible that a recessive or a dominant gene is causing the problems in this family. A detailed family history, therefore, should be taken, looking for evidence of congenital cardiac malformations. The role of parental echocardiography in risk assessment is unclear. It may identify asymptomatic patients with congenital cardiac anomalies, suggestive of dominant inheritance. It may also reveal anatomical variants, the significance of which is uncertain, such as an aortic valve with two leaflets. A couple in this situation has a risk of recurrence of any type of congenital cardiac anomaly of between 5% and 10%. This takes account of all of the possible causes of hypoplasia of the left heart. Detailed fetal echocardiography will be offered.

Offering Advice on the Risk of Recurrence

Assessment of the risk of recurrence is based on the family history; clinical examination of the child, which may identify a specific syndrome; and the results of investigations, such as chromosomal analysis or the postmortem report.

For most couples, an empirical risk of recurrence will be offered. There are possible problems and pitfalls (Figs. 4.1–4.3) that highlight the potential demand for specific investigations to identify families with a higher risk of recurrence.

Acknowledgments

We wish to thank the families that have taken part in our studies, and the support of Little Hearts Matter and the Emily Hey Fund. We are also very grateful to the British Heart Foundation, which funds our research.

References

1. Nora JJ, Nora AH. The genetic contribution to CHD. In: Nora JJ, Takao A, eds. CHD: Causes and Processes. London: Futura, 1984.

2. Boughman JI, Berg KA, Astemborski JA, et al. Familial risks of congenital heart defect assessed in a population-based epidemiologic study. Am J Med Genet 1987;26: 839–849.
3. Shokeir MHK. Hypoplastic left heart syndrome: an autosomal recessive disorder Clin Genet 1971;2:7–14.
4. Gerboni S, Sabatino G, Mingarelli R, Dallapiccola B. Coarctation of the aorta, interrupted aortic arch, and hypoplastic left heart syndrome in three generations. J Med Genet 1993;30:328–329.
5. Brenner JA, Berg KA, Schneider DS, Clark EB, Boughman JA. Cardiac malformations in relatives of infants with hypoplastic left heart syndrome. Am J Dis Child 1989;143:1492–1495.
6. Schinzel AA. Cardiovascular defects associated with chromosomal aberrations and malformation syndromes. Prog Med Genet 1983;5:303–381.
7. Brewer C, Holloway S, Zawalnyski P, Schinzel A, Fitzpatrick D. A chromosomal deletion map of human malformations. Am J Hum Genet 1998;63:1153–1159.
8. Van Karnebeek CDM, Hennekam R. Associations between chromosomal anomalies and congenital heart defects: a database search. Am J Med Genet 1999; 84:158–166.
9. Phillips H, Renforth GL, Spalluto C, et al. Narrowing the critical region within 11q24-qter for hypoplastic left heart and identification of a candidate gene JAM3, expressed during cardiogenesis. Genomics 2002;79:475–478.
10. Andrews R, Tulloh R, Sharland G, et al. Outcome of staged reconstructive surgery for hypoplastic left heart following antenatal diagnosis. Arch Dis Child 2001;85: 450–451.
11. Brackley KJ, Kilby MD, Wright JG, et al. Outcome after prenatal diagnosis of hypoplastic left-heart syndrome: a case series. Lancet 2000;356:1143–1147.
12. van Egmond H, Orye E, Praet M, Coppens M, Devloo-Blancquaert A. Hypoplastic left heart syndrome and 45X karyotype. Br Heart J 1988;60:69–71.
13. Natowicz M, Chatten J, Clancy R, et al. Genetic disorders and major extracardiac anomalies associated with the hypoplastic left heart syndrome. Pediatrics 1988;82: 698–706.
14. Johnson VP, Petersen LP, Holzwarth DR. Midtrimester prenatal diagnosis of short-limb dwarfism (Salido-Noonan syndrome). Birth Defects 1982;18:133–141.
15. Kirk E, Ades L. Hypoplastic left heart in cerebrocostomandibular syndrome. J Med Genet 1998;35:879.
16. Greenwood RD, Rosenthal A, Parisi L. Extracardiac anomalies in children with congenital heart disease. Pediatrics 1975;55:485–492.
17. Hall BD. Choanal atresia and associated multiple anomalies. J Pediatr 1979;95 395–398.
18. Cryan SE, Martinez R, Daniels. Spectrum of congenital heart disease in CHARGE association. J Pediatr 1987; 110:576–578.
19. Day-Salvatore D, McLean D. Blepharophimosis, hypoplastic radius, hypoplastic left heart, telecanthus, hydronephrosis, fused metacarpals and "prehensile" halluces: a new syndrome? Am J Med Genet 1998;80: 309–313.
20. Gianotti A, Diglio MC, Mingarelli R, Dallapiccola B. An autosomal recessive condition of cleft palate, cardiac defect, genital anomalies, and ectrodactyly (CCGE). J Med Genet 1995;32:72–74.
21. Holzgreve W, Wagner H, Rehder H. Bilateral renal agenesis with Potter phenotype, cleft palate, anomalies of the cardiovascular system, skeletal anomalies including hexadactyly and bifid metacarpal. A new syndrome? Am J Med Genet 1984;18:177–182.
22. Hurst JA, Hallam LA, Hockey KA. Leptomeningeal angiomatosis, absent olfactory tracts, hypoplasia of the cerebellar vermis, cleft lip and palate and congenital heart disease in a stillborn infant. Clin Dysmorphol 1992; 1:168–171.
23. Kaufman RL, Hartmann AF, McAlister WH. The association of severe obstructive left heart lesions, vertebral and renal anomalies in sibs (family studies in congenital heart disease I). Birth Defects 1972;8(5):82–84.
24. Kennerknecht I, von Saurma P, Brenner R, et al. Agonadism in two sisters with XY gonosomal constitution, mental retardation, short stature, severely retarded bone age, and multiple malformations: a new autosomal recessive syndrome. Am J Med Genet 1995;59:62–67.
25. McPherson E, Clemens M. Cleft lip and palate, characteristic facial appearance, malrotation of the intestine, and lethal congenital heart disease in two sibs: a new autosomal recessive condition? Am J Med Genet 1996;62:58–60.
26. Medeira A, Dennis N, Donnai D. Anencephaly with spinal dysraphism, cleft lip and palate and limb reduction defects. Clin Dysmorphol 1994;3:270–275.
27. Neish AS, Roberts DJ. Hypoplastic left heart, nephromegaly, and distinctive facies in two siblings (Abstract). Am J Hum Genet 1992;50:153.
28. Newbury-Ecob RA, Zuccollo JM, Rutter N, Young ID. Sex linked valvular dysplasia. J Med Genet 1993;30: 873–874.
29. Bartsch O, Wagner A, Hinkel GK, et al. FISH studies in 45 patients with Rubinstein-Taybi syndrome: deletions associated with polysplenia, hypoplastic left heart and death in infancy. Eur J Hum Genet 1999;7:748–756.
30. Hannauer D, Argilla M, Wallerstein R. Rubinstein-Taybi syndrome and hypoplastic left heart. Am J Med Genet 2002;112:109–111.
31. Czarnecki P, Lacombe D, Weiss L. Toriello-Carey syndrome: evidence for X-linked inheritance. Am J Med Genet 1996;65:291–294.

Chapter 5

DIAGNOSTIC APPROACH

James L. Wilkinson

Hypoplasia of the left heart encompasses a group of conditions that are all characterised by moderate or severe hypoplasia or atresia of the mitral valve, left ventricle, and aortic valve, and variable hypoplasia of the ascending aorta. The left atrium may also be small, but this is less constant. Frequently the aortic valve is imperforate, producing aortic atresia, with tiny sinuses of Valsalva and totally fused commissures. The ascending aorta serves merely to carry retrograde flow from the aortic arch to feed the coronary arteries (Fig. 5.1), and may be little larger than they are. In such cases, the entire systemic circulation is dependent on flow through the arterial duct from the pulmonary trunk and morphologically right ventricle. The left ventricle is usually a tiny and thick-walled cavity, often with endocardial fibroelastosis. The mitral valve is diminutive and frequently severely abnormal, with thickened, dysplastic, leaflets and short tendinous cords. The left atrium is usually thick walled, and may also show endocardial fibroelastosis. Its only functional exit is via the oval foramen, or an atrial septal defect, to the right atrium. Associated coarctation is frequently present, though seldom of much clinical importance at the time of presentation.

In other variants hypoplasia of the left heart, both aortic and mitral valves may be patent but hypoplastic and/or stenotic. Alternatively, both mitral and aortic valves may be atretic. The left ventricle is then merely a slit in the posterior ventricular wall. In some cases, either with mitral or aortic atresia, a ventricular septal defect may be present. In these instances, the left ventricle, though still small, may be better represented. Rarely there may be a common atrioventricular junction with an atrioventricular septal defect, and it is then the left component of the common atrioventricular valve that will be hypoplastic, as well as the left ventricle and the aorta.

In another related malformation, both great arteries can arise from the right ventricle, with the ventricular septal defect in the subaortic position, and again with hypoplasia of the left ventricle and ascending aorta. In all these variants, an associated coarctation is frequent, most usually in the preductal position.

Hypoplasia of the left heart as a constellation of lesions, however, needs to be distinguished from those examples of coarctation-associated aortic valvular stenosis but with less severe hypoplasia of the components of the left heart. In the latter group of patients, the hallmark of the anomaly is that the ascending aorta is larger than the arch or isthmic segment, whilst in those with true hypoplasia of the left heart the reverse is usually true (cf. Fig. 5.1 and Fig. 5.2B).

It should be noted that aortic stenosis or atresia with hypoplasia of the ascending aorta can occur in association with more complex defects, such as discordant ventriculoarterial connections, or functionally univentricular hearts. These cases are not usually categorised as hypoplasia of the left heart, albeit that the clinical presentation may be similar.

There is then another subgroup of infants who manifest some features of both coarctation and hypoplasia of the left heart. In many such cases, the degree of left ventricular hypoplasia is less severe, although the mitral and aortic valves and the left ventricle are all small. These infants frequently exhibit evidence of multiple obstructive problems within the left heart, with subaortic and/or aortic valvular stenosis, mitral and/or supramitral stenosis, and coarctation, as well as some degree of generalised left ventricular hypoplasia. The term "Shone's complex"[1] has been applied to such

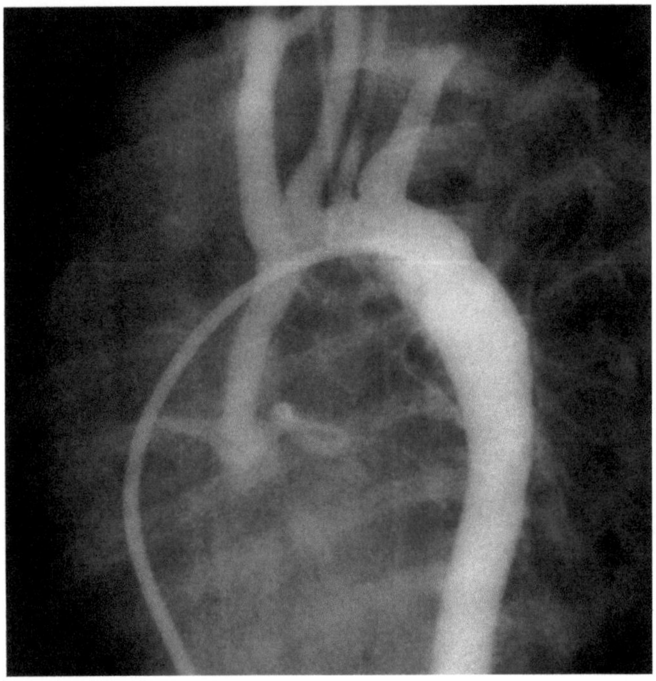

Figure 5.1. Angiogram showing descending aorta, filled via injection through the arterial duct, with retrograde filling of the ascending aorta and coronary arteries. Note that the ascending aorta is smaller than the arch and descending aorta.

patients, and they present a major clinical and surgical challenge!

HAEMODYNAMICS

In almost all variants of hypoplasia of the left heart, all or most of the pulmonary venous blood is diverted at the atrial level through an interatrial communication to the right atrium. As the flow of blood is frequently obstructed by a narrow oval foramen, pulmonary venous hypertension is usual. The right atrium and right ventricle are greatly dilated, and pulmonary arterial pressure is invariably high, and may be suprasystemic. Pressure and flow in the aorta are dependent on the state of ductal constriction. If widely patent, the systemic pressure and perfusion may be normal, but as the duct closes there will be a progressive fall in aortic pressure, until generalised acidosis and myocardial hypoxia and hypoperfusion, with progressive myocardial impairment, supervene.

CLINICAL FEATURES

In the early hours after delivery, the infant with hypoplasia of the left heart may appear healthy. Peripheral pulses may be palpable, and obvious cyanosis may be absent. Tachypnoea and variable cyanosis tend to develop within hours of days of birth, and a prominent parasternal cardiac impulse, with or without a soft systolic murmur, is then present. The second sound is single. After a period of hours, or a few days, the peripheral pulses become diminished or absent, as the duct closes, and manifestations of cardiac failure appear. Hepatomegaly is usually striking. Progression is rapid. Within a few hours from the onset of symptoms, the infant is often moribund, and in a state of circulatory collapse.

Whilst some patients present with classical signs and symptoms within hours of birth, others may remain relatively well, feeding normally for several days, and presentation may be delayed until the end of the first week of life, or, rarely, even later.

Distinction from the clinical picture of the coarctation constellation is sometimes difficult, but it should be noted that, in those with hypoplasia of the left heart, there is no discrepancy between the pulses or pressures in the upper and lower parts of the body, as the entire systemic circulation is more or less equally compromised concomitant with closure of the arterial duct.

In infants with features of coarctation, who also have features of "borderline" hypoplasia of the left heart, the degree of left ventricle hypoplasia is often less severe, and the circulation to the upper body may be dependent on the left ventricle, in which case the pulses in the arms may be much more readily palpable than those in the legs.

INVESTIGATIONS

Establishment of the diagnosis is a matter of urgency, as many affected infants are extremely ill by the time they present for medical attention. It is almost always possible, nonetheless, to stabilise the condition of such babies by a combination of intravenous prostaglandin E, inotropic agents such as dopamine, correction of acid/base metabolism, expansion of volume, and positive pressure ventilation. Such measures can lead to dramatic improvement in the cardiovascular state, and usually should be instituted as soon as possible. It is undesirable to withhold treatment until after the diagnosis has been made, if this is likely to produce any significant delay.

Chest X-Ray

The x-ray usually shows marked cardiomegaly and pulmonary congestion. The appearances are nonspecific, however, and do not distinguish hypoplasia of the left heart from other malformations that may present in the

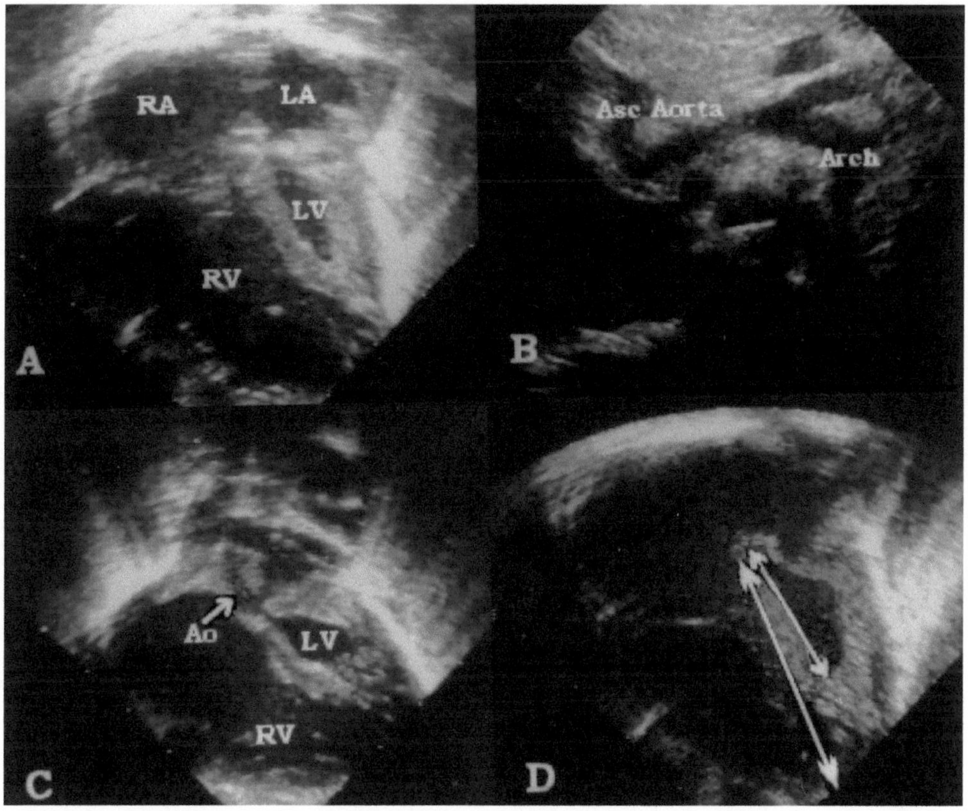

FIGURE 5.2. Echocardiographic frames from infants with hypoplasia of the left heart. A: Four-chamber view showing severe hypoplasia of the left ventricle. B: View of the aortic arch demonstrating the ascending aorta, which measures approximately 3 mm in diameter, and which is smaller than the arch. C: Four-chamber and aortic root view showing gross hypoplasia of the left ventricle and aortic root. D: Arrows demonstrate the measurements taken to calculate the ratio of the left-to-right ventricular long axis. Measurements are made from the endocardial surface of the apex of each ventricle to the hinge points of the mitral or tricuspid valve at the crux.

newborn period with similar symptoms, including the coarctation constellation and interruption of the aortic arch. Less commonly, the heart may be normal in size, with lung fields that manifest intense pulmonary venous congestion due to severe pulmonary venous obstruction, often resulting from premature closure of the oval foramen.

Electrocardiogram

The electrocardiogram shows evidence of reduced or absent left ventricular activity with right axis deviation, right ventricular, and sometimes right atrial hypertrophy. The normal small septal Q wave, seen in the left chest leads V_5 and V_6, is characteristically absent (Fig. 5.3). In those with dilated right ventricles, however, there can be marked clockwise rotation, and well-developed R waves may sometimes be seen in the lateral chest leads, albeit without septal Q waves. Superior, or left, axis deviation may be present in infants with the right dominant form of atrioventricular septal defect, in some of whom severe hypoplasia of the structures of the left heart justifies inclusion within the group under discussion, though this variant of atrioventricular septal defect is not always associated with the typical electrocardiographic findings, namely left axis deviation and right bundle branch block.

Echocardiography

Differentiation from other causes of severe cardiac failure or shock syndrome needs to be made rapidly. The most useful diagnostic tool is the echocardiogram, with which the degree of hypoplasia of the left ventricle and aortic root can be assessed rapidly, precisely, and noninvasively (Fig. 5.2). The anatomy of the left ventricle, and of the mitral and aortic valves, may be studied from the parasternal long axis view, and from the four-chamber view, either from the apex or from the subziphoid window. Additional information is obtainable from the parasternal short axis and from suprasternal cuts, especially in relation to the aortic arch and duct. Doppler interrogation and colour flow mapping will potentially clarify questions about forward flow through a tightly stenotic aortic valve, and will allow assessment of mitral and tricuspid regurgitation.

Differentiation of hypoplasia of the left heart from other cardiac defects with a duct-dependent systemic circulation, for example, coarctation, aortic interruption, or critical aortic stenosis, is usually straightforward. As already discussed, nonetheless, some infants may manifest features of aortic stenosis or coarctation and of typical hypoplasia of the left heart, often with less severe hypoplasia of the left-sided structures. This group of patients present major difficulties from the point of

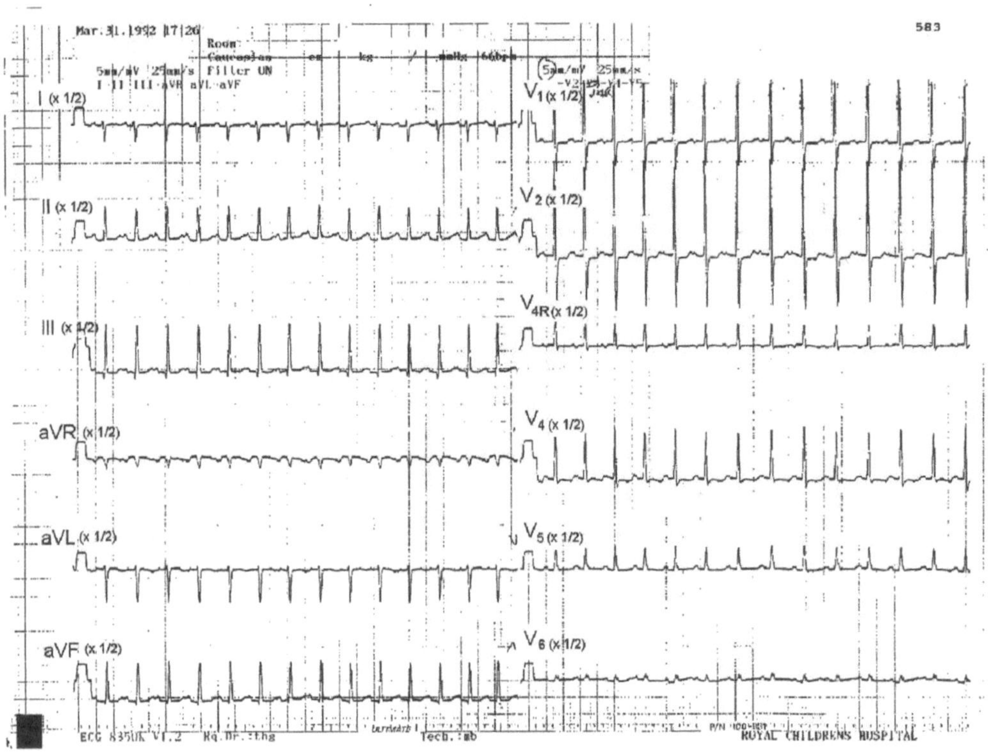

FIGURE 5.3. Electrocardiogram from an infant with hypoplastia of the left heart. There are voltage changes of right ventricular hypertrophy, with reduced left ventricular voltages in V_6 and absent septal Q waves in left ventricle leads. All leads recorded on 6 mm/mV standardisation.

view of deciding the appropriate form of treatment (see below).

ANTENATAL DIAGNOSIS

Antenatal diagnosis of hypoplasia of the left heart may be achieved, with fetal ultrasound, as early as 16 to 18 weeks' gestation (Fig. 5.4). In most cases, the diagnosis is suspected in the course of an examination done for noncardiac reasons. Often a routine fetal scan reveals, in the four-chamber view, asymmetric development of the ventricles as the first clue of the presence of an abnormality.[2,3] This finding demands urgent review by an experienced fetal echocardiographer. In some cases, the diagnosis can be made with confidence at this stage, and counselling, with the possibility of late termination of pregnancy, may proceed.[2] In other cases, however, the abnormality may be much less apparent at this stage of gestation, and the full-blown picture of hypoplasia of the left heart may not evolve until later, often too late for termination of pregnancy. There are an increasing number of reports from experienced fetal echocardiographers of examples in which a relatively well-formed left ventricle, usually with evidence of aortic stenosis, was demonstrated at 18 to 20 weeks' gestation, and subsequently failed to grow as the pregnancy progressed so that, by 30 to 36 weeks' gestation, the ventricle had become severely hypoplastic.[4] At the same time the right ventricle, which carries the large majority of the fetal cardiac output in such infants, becomes progressively dilated so that the asymmetry between the two is exaggerated.

The major benefits of early diagnosis in fetal life are that, at least when the diagnosis is made before 20 weeks, the option for termination of pregnancy exists and that, regardless of the stage when the diagnosis is made, counselling of the parents regarding the options for

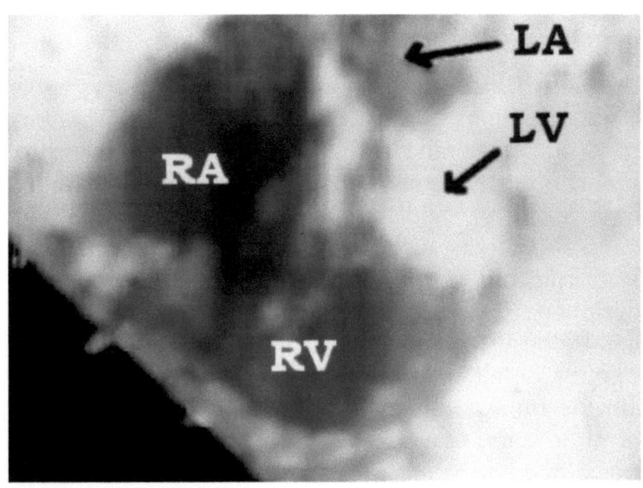

FIGURE 5.4. Frame of a fetal echocardiogram, at 18 weeks' gestation, in four-chamber view, showing gross hypoplasia of the left ventricle with obliteration of the cavity.

treatment may take place prior to delivery, and a plan for management can be made. Because of fetal diagnosis and selective termination, the frequency with which hypoplasia of the left heart presents at or after delivery has fallen in many centres.[5] The rate of detection of the malformation in pregnancies in which a scan has been performed, nonetheless, remains suboptimal, being barely more than one-third in one study.[6] Most affected infants continue to remain undetected until after birth. It is also noteworthy that asymmetric development of the ventricles, leading to a suspicion of hypoplasia of the left heart on fetal echocardiography, may occasionally, in the absence of other, more definite, abnormalities, persist without progression as gestation advances, with postnatal evolution to normal cardiac function. Thus, the presence of an apparently small, but otherwise normal, left heart on an early fetal scan should be interpreted with caution, especially when the degree of hypoplasia is not severe.

"BORDERLINE" HYPOPLASIA OF THE LEFT HEART

A number of patients manifest degrees of hypoplasia of left heart structures, including the aortic and mitral valves and left ventricle, but may have sufficiently well-formed valves and/or left ventricles to leave serious doubt as to whether the usual surgical therapy for classical hypoplasia, such as the Norwood cascade or transplantation, is appropriate. In many such cases, the dominant lesion is aortic stenosis, with or without associated coarctation. The option of balloon aortic valvoplasty, or surgical aortic valvotomy, is clearly preferable to a Norwood procedure, especially when there is a reasonable prospect of achieving survival with a biventricular circulation. Until recently, most centres would have given such infants the benefit of the doubt. Unfortunately, such borderline cases present a major challenge. All too often, in the past, the temptation to attempt either surgical or balloon valvotomy, rather than to elect for a Norwood operation or transplantation, has been followed by the death of the patient, either perioperatively or later, sometimes after a prolonged period of illness and multiple procedures. There is clearly a need to define which patients with hypoplasia of the left-sided structures stand a good chance of surviving after conservative procedures such as aortic valvotomy, as opposed to those who will not benefit from such treatment, and therefore should be managed with a Norwood procedure or transplantation.

Many centres have evolved local policies for making decisions for this group of patients. A range of factors and measurements has been used in an effort to allow accurate prediction of the likelihood of survival to be achieved. These include the size of the left ventricular cavity, the presence of endocardial fibroelastosis,[7] aortic orificial diameter,[8,9] and haemodynamic parameters.[8] There is general agreement that a diameter of the aortic root of less than 5 mm in a newborn infant, or a mitral valvar annular diameter of less than 8 mm, especially when the apex of the left ventricle does not form the cardiac apex, is strongly indicative of a poor outlook should conservative surgery be attempted. Patients with such features, therefore, tend to be considered unsuitable for valvotomy or valvoplasty. On the other hand, the size of the left ventricular cavity correlates less well with survival.[8,10]

A useful study from Boston has sought to add an element of precision to this decision-making process.[11] These investigators analysed a cohort of infants with aortic stenosis and varying degrees of hypoplasia of the left heart who had been managed with valvotomy or valvoplasty. The authors analysed which measurements, from a preoperative echocardiogram, were independent variables predicting the likelihood of nonsurvival. The measurements included:

- Aortic orificial diameter
- Aortic root
- Left ventricular cavity
- Mitral annulus and area
- Left ventricular mass
- Ratio of left and right ventricular long axis

Of these, the most useful proved to be the diameter of the aortic root, indexed to body surface area; the area of the mitral valve, also indexed, and calculated by using the annular diameter in the long axis plan and that in the four-chamber view (Fig. 5.5); and the ratio of the left and right ventricular axes (Fig. 5.2D).

The threshold values recommended for attempted biventricular repair were

- Mitral valvar area greater than 4.75 cm^2/m^2
- Diameter of aortic root greater than 3.5 cm^2/m^2
- Ratio of left to right ventricular long axis greater than 0.8

The presence, in any individual case, of values equal to or less than these, for two such measurements, was predictive of death in all cases. If only one of the three was at or below threshold, survival was probable. It seems clear, therefore, that any patient in whom two or more of these measurements fall below the above values should be considered unsuitable for such therapy, and should be managed with a Norwood procedure or

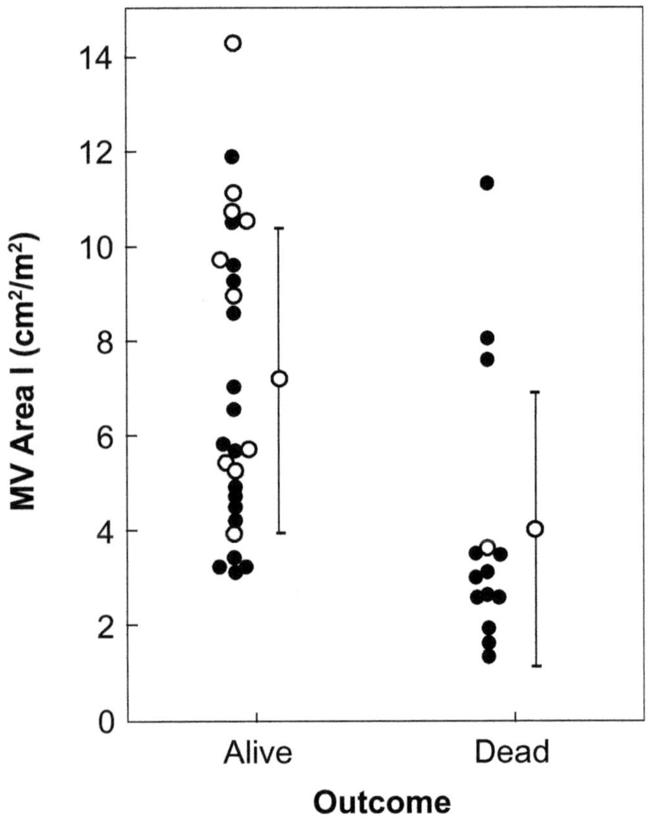

FIGURE 5.5. Scatterplot of the mitral valvular area, indexed to body surface area, measured from echocardiograms, in infants with aortic stenosis who had aortic valvotomy or valvoplasty. Most of those with smaller mitral valvular areas died, while almost all survivors had larger areas. Similar plots can be produced for the other variables measured. (From Rhodes et al.,[11] with permission.)

transplant. It is noteworthy that left ventricular volume was not a particularly useful predictor of death. Left ventricular mass, nonetheless, when taken in conjunction with other variables, did improve the accuracy of prediction of death or survival. Thus, the presence of an apparently adequate left ventricular cavity is of limited value in determining the likelihood of survival after valvotomy or valvoplasty.

A significant typographical error occurred in the Boston paper that can cause confusion, and should be noted by any who may wish to make calculations of the mitral valvar area according to their recommendations.[11] The formula for calculation of this area is given as

$$\text{Area} = \frac{([D1 \times D2])}{2}$$

where $D1$ is the long axis diameter, and $D2$ is the 4-chambers diameter.

The correct formula for the area of an ellipse is

$$\text{Area} = \frac{(D1 \times D2)}{4}$$

The formula that appears in the paper would lead to values that are double the real figure. The measurements used and tabulated (Fig. 5.5) were made using the correct formula, so that the error does not affect the validity of the observations or the recommendations.[11] Those who choose to calculate the mitral valvar area according to the recommendations of the Boston group,[11] however, should be aware of this error and employ the proper formula.

In practice, as newborn infants with classical hypoplasia of the left heart usually weigh in the range of 2.5 to 3 kg, with a surface area close to 0.2 m^2, the following measurements may be regarded as threshold values:

- Mitral valvar diameter greater than 1.1 cm, and area greater than 0.95 cm^2 (Fig. 5.6)
- Diameter of aortic root greater than 7 mm

It should be noted that the measurement of the aortic root is made at the level of the aortic sinuses, not at the "annulus" (Fig. 5.6). The equivalent measurement for the so-called annulus is smaller, at approximately 5 mm. The threshold diameter for the annulus of the mitral valve, according to the Boston data, is clearly larger than the value employed by many groups, which is generally about 8 mm. The aortic dimensions, with an annulus of 5 mm and a root of 7 mm, are broadly similar to those arrived at in other centres.

CONCLUSION

The diagnosis of hypoplasia of the left heart has become relatively straightforward with the advent of good-quality cross-sectional echocardiography. Antenatal diagnosis is feasible in some cases from about 18 weeks, though not all affected fetuses show typical features that early in gestation. Prenatal diagnosis allows consideration of termination of pregnancy, and also permits counselling and early decisions about postnatal management.

Borderline cases, presenting postnatally, can pose difficult problems in reaching a decision about management. Criteria have been derived that allow accurate predictions to be made about the likelihood of death after "conventional" surgery, such as aortic valvotomy or valvoplasty, in patients with less severe degrees of hypoplasia of the left-sided structures, and in whom

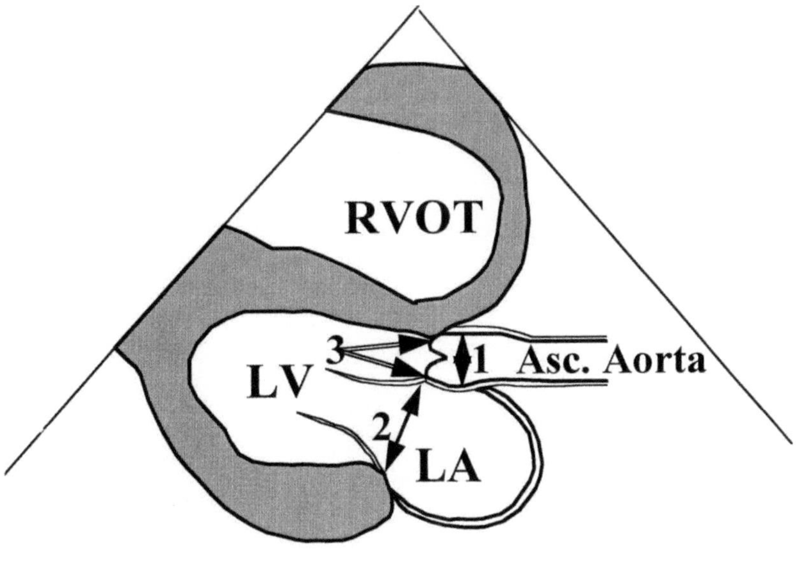

FIGURE 5.6. Diagrammatic representation of an echocardiogram, in parasternal long axis view, in an infant with moderately hypoplastic left heart structures. The sites of measurement of the aortic root diameter (1) and the mitral annulus diameter (2) are indicated. The aortic (3) dimension, taken at the hinge point of the leaflets, is smaller than the dimension of the root (1).

aortic stenosis is a major intracardiac defect. Patients in whom the measurements of the aortic root, mitral valve, and the left-to-right ventricular ratio fall below the cut-off levels given above, for two of the three measurements, should be regarded as unsuitable for such conventional surgery. For such infants, the only prospect for survival is likely to be a Norwood operation or, in those centres where it is possible, neonatal transplantation.

These criteria should be used with caution in infants in whom the dominating problem is not aortic stenosis, such as those in whom coarctation is the main anomaly. Unfortunately there are no equivalent data for infants with borderline hypoplasia of the left-sided structures in whom the immediate problem requiring intervention or surgery is not aortic stenosis.

References

1. Shone JD, Sellers RD, Anderson RC, Adams P, Lillehei CW, Edwards JE. The developmental complex of "parachute mitral valve," supravalvular ring of left atrium subaortic stenosis and coarctation of the aorta. Am J Cardiol 1963;11:714–724.
2. Allan LD, Cook A, Sullivan I, Sharland GK. Hypoplastic left heart syndrome: effects of foetal echocardiography on birth prevalence. Lancet 1991;337:959–961.
3. McGahan JP, Choy M, Parrish MD, Brandt WE. Sonographic spectrum of foetal cardiac hypoplasia. J Ultrasound Med 1991;10:539–546.
4. Sharland GK, Chita SK, Fagg NL, et al. Left ventricular dysfunction in the foetus: relation to aortic valve anomalies and endocardial fibroelastosis. Br Heart J 1991;66: 419–424.
5. Allan LD. Foetal diagnosis of fatal congenital heart disease. J Heart Lung Transplant 1993;12(6 pt 2):S159–160.
6. Stoll C, Alembik Y, Dott B, Roth PM, De Geeter B. Evaluation of prenatal diagnosis of congenital heart disease. Prenatal Diagnosis 1993;13:453–461.
7. Morcellin R, Sauer U, Simon B, Comazzi M, Sebening F, Buhlmeyer K. Reduced left ventricular cavity size and endocardial fibroelastosis as correlates of mortality in newborns and young infants with severe aortic valve stenosis. Paediatr Cardiol 1983;4:265–272.
8. Pelech AN, Dyck JD, Trusler GA, et al. Critical aortic stenosis: survival and management. J Thorac Cardiovasc Surg 1987;94:510–517.
9. Turley K, Bove EL, Amato JJ, et al. Neonatal aortic stenosis. J Thoac Cardiovasc Surg 1990;99:679–684.
10. Gundry SR, Behrendt DM. Prognostic factors in valvotomy for critical aortic stenosis in infancy. J Thorac Cardiovasc Surg 1986;92:747–754.
11. Rhodes LA, Colan S, Perry SB, Jonas RA, Sanders SP. Predictors of survival in neonates with critical aortic stenosis. Circulation 1991;84:2325–2335.

Chapter 6

PRE- AND POSTOPERATIVE MANAGEMENT OF INFANTS WITH HYPOPLASIA OF THE LEFT HEART

Thomas J. Kulik, Dennis C. Crowley, and John R. Charpie

A newborn with hypoplasia of the left heart is much more likely to survive surgical palliation today than 10 years ago for two reasons: better surgery and better preoperative, and especially postoperative, care. Over the last decade, surgeons have learned how to avoid obstruction within the aortic arch and coronary arteries, and have developed optimal strategies for constructing the systemic-to-pulmonary arterial shunt. At the same time, a better appreciation of the dynamics of the pulmonary and systemic circulations has evolved, as have more effective means of controlling these and other factors crucial to a good outcome. There is much yet to be learned. For example, we still lack a good means of selectively increasing pulmonary vascular resistance, which would be of great benefit for some infants following Norwood palliation. Our understanding of how the multiple insults that attend this condition, including cardiopulmonary bypass, ventricular volume overload, reduced coronary arterial perfusion, and so on, affect myocardial biochemistry and function. Our appreciation of how to influence these factors remains extremely rudimentary. Operative survival in excess of 90% now achieved in some centres (see Chapter 7) suggests, however, that we know a lot more than we did even a few years ago.

This chapter summarises our current approach to the pre- and postoperative management of newborns with hypoplasia of the left heart who undergo Norwood palliation. Our experience, however, may not readily be translatable to other institutions, especially if taken in bits and pieces, where other details of surgical and medical management may differ from ours. Also, most of our approach has evolved via impressions formed while taking care of patients, rather than through rigorous, prospective studies. We hope that these imperfections do not preclude its usefulness for those who care for this group of infants.

PREOPERATIVE MANAGEMENT

Initial Cardiac Evaluation

Physical Examination

As is true for any infant, the physical exam is the initial element of the assessment. The classic physical findings of hypoplasia of the left heart are described elsewhere in the book (see Chapter 5), but it should be noted here that the physical examination in these babies should be aimed both at discovering any associated congenital abnormalities and at estimating the systemic perfusion. Patients palliated with intravenous prostaglandin E_1 are generally reasonably warm, with brisk capillary refill, and excellent or even bounding pulses. Poor pulses and diminished peripheral perfusion suggest one or more problems, for example, hypovolemia, excess flow of blood to the lungs, inadvertent interruption of the prostaglandin, sepsis, or hypothermia. These problems need resolution.

Echocardiography

Several pieces of information regarding cardiac anatomy and physiology are important in deciding the most appropriate therapy for infants with left-sided obstructive lesions. The first decision to be made is whether the cardiac anatomy is appropriate for a biventricular repair, for example, when critical aortic stenosis is associated with an acceptable, well-developed, and good-sized left ventricle, mitral valve, and aortic valve, or whether only one functioning ventricle can be provided. The size and function of all four cardiac valves, especially the aortic and mitral valves, the size and systolic function of the left, if present, and right ventricles, and the size and anatomy of the aortic arch can all be determined using cross-sectional echocardiography supplemented with Doppler. Indeed, echocardiography has largely supplanted cardiac catheterisation in establishing these features. Besides supplying information necessary to plan for uni- or biventricular surgical approach, echo/Doppler studies provide other useful information. Thus, most babies with hypoplasia of the left heart who have not suffered significant cardiovascular collapse have qualitatively normal, or perhaps slightly depressed, right ventricular systolic function at presentation. Echocardiographic evidence of significantly depressed right ventricular systolic function may indicate the need for inotropic support. For those who present in shock, these investigations indicate the extent of the cardiovascular compromise, and show when sufficient recovery has occurred to permit the operation to proceed. While the relationship between the size of the ascending aorta and survival with Norwood palliation has yet to be established, in some centres the finding of a very small ascending aorta might be considered an indication for an approach other than Norwood palliation. Echocardiographic evidence of significant tricuspid valvular regurgitation might similarly modify the medical or surgical approach to the patient, especially if other risk factors are present. Abnormal pulmonary and systemic venous connections can occur, and it is particularly important to diagnose and define any anomalous pulmonary venous connections prior to embarking on Norwood palliation. A small pressure gradient across the atrial septum, of less than about 7 mm Hg, is not uncommon, and does not presage high pulmonary vascular resistance in the postoperative period. Higher pressure gradients can result in hypoxemia and pulmonary oedema preoperatively, and severe hypoxia postoperatively. We discuss the management of the patient with a restrictive atrial septal communication below.

Chest Radiograph

The chest radiograph is useful in suggesting the magnitude of the flow of blood to the lungs, which can be assessed from the prominence of the vascular markings, and in assessing the presence of interstitial oedema, and cardiac size. It also indicates the presence of pulmonary venous hypertension through the interstitial oedema. Pulmonary lymphangectasia has been described in babies with hypoplastic of the left heart (see Chapter 2), and obstruction to pulmonary venous inflow may also be apparent on the chest x-ray.

Electrocardiography

The 12-lead electrocardiography usually supplies little additional information, although occasionally there is evidence of ventricular ischemia, resulting from either a period of low perfusion, or very rarely insufficient coronary arterial flow due to obstruction of the aortic arch proximal to the insertion of the arterial duct. Arrhythmias and manifest accessory atrioventricular muscular connections are uncommon in these patients, although we have observed a few of these babies to have supraventricular tachycardia, presumably due to a concealed bypass track.

Arterial Blood Gases

Arterial blood gases are important in the initial evaluation of these patients. While a mildly depressed level of bicarbonate in the serum may occur in the otherwise normal neonate, levels below 18 suggest that systemic perfusion may be inadequte, which requires evaluation and perhaps specific therapy (see below). The arterial saturation of oxygen is also a very important variable, both initially and throughout the preoperative course (see below).

Initial Noncardiac Evaluation

It is important to evaluate these babies for noncardiac abnormalities. Measurement of electrolytes, hematocrit, blood urea, and nitrogen is indicated in all such neonates. For those who present with shock, measures of hepatic function, the coagulative state, and cranial ultrasonography to rule out haemorrhage are also indicated.

Routine Preoperative Medical Management

Maintenance of Ductal Patency

The initial dose of intravenous prostaglandin E_1 needed to open a closed or closing arterial duct is 0.1 μg/kg/min. To maintain patency of the open duct, we use 0.3 mg/kg/min. It should be noted that this drug is a vasodilator, and therefore it decreases filling pressure and blood pressure, especially when first started. Volume infusion, therefore, is often necessary when starting prostaglandin E_1.

Indications for Mechanical Ventilation

There is nothing in the anatomy or physiology of neonates with hypoplasia of the left heart that mandates that all such patients be mechanically ventilated. In fact, a significant percentage of these babies can provide their own ventilation until surgery, and there are several advantages to avoiding intubation, such as reduced risk of airway trauma, reduced need for sedation, and lesser need for intensive monitoring. The chief indications for intubation are apnoea, almost always related to infusion of prostaglandins, and substantially increased work of breathing, especially if accompanied by findings of diminished systemic perfusion. Patients with a significantly compromised cardiovascular state, for example, shock secondary to ductal closure, or severe hypoxemia due to a highly restrictive atrial septal opening, also require intubation and mechanical ventilation. We have rarely found it necessary to intubate a patient for the purpose of reducing the minute ventilation to increase the arterial tension of carbon dioxide, although this is a consideration in an occasional patient.

Securing an Appropriate Ratio of Pulmonary-to-Systemic Flow

The ratio of pulmonary-to-systemic flow in these patients is largely dependent on the ratio of the pulmonary to the systemic vascular resistance. Because pulmonary vascular resistance normally falls to less than one-quarter of systemic vascular resistance in the first few days of life, there is a tendency for the greater part of the output from the right ventricle to find its way into the lungs, with undesirable consequences. The right ventricle suffers a large volume load, systemic blood flow is reduced, and because of diastolic runoff into the lungs, coronary arterial perfusion is impaired.

At any given ventricular output, be it from a functionally single or two ventricles, every millilitre of blood pumped to the lungs is one less millilitre that goes to the systemic circulation. As the ratio of pulmonary to systemic flow increases, therefore, and as the systemic saturation of arterial oxygen increases, the systemic flow falls. Thus, systemic transport of oxygen is a function of the ratio of pulmonary to systemic flows itself. The relationship between the ratio of flows and systemic transport of oxygen is sigmoidal[1] (Fig. 6.1). The theoretical analysis of Barnea and colleagues[1] suggests that the optimal arterial saturation of oxygen in this situation, relative to systemic delivery of oxygen, is between 60% and 70%.

While excess pulmonary relative to systemic flow is the most common situation in these babies, too little pulmonary flow, and severe hypoxemia, is occasionally encountered. Flow of blood to the lungs may be reduced due to increased pulmonary vascular resistance, which may be idiopathic, as is seen in persistent pulmonary hypertension of the newborn. Increased resistance to flow can also be caused by pulmonary venous hypertension; alveolar hypoxemia; acidosis, be it respiratory or metabolic; or lung disease.

Although the ratio of pulmonary to systemic flow is a key determinant of the arterial saturation of oxygen, other factors are also important. Pulmonary venous

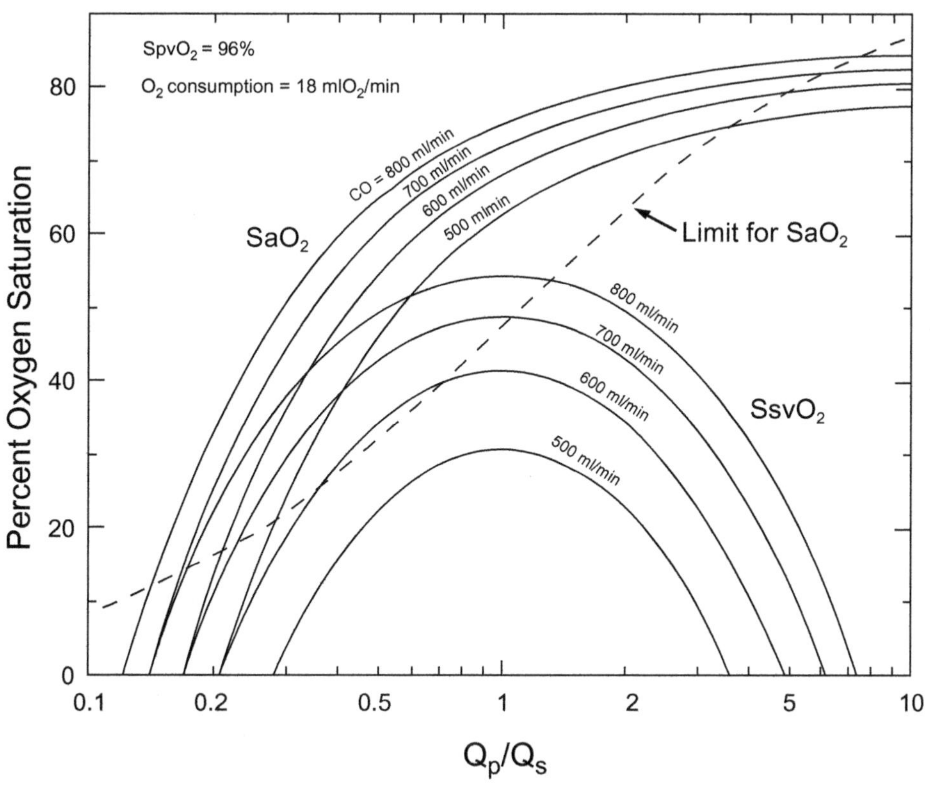

FIGURE 6.1. A mathematical model was developed to relate the ratio of pulmonary-to-systemic flows (Q_p/Q_s), systemic arterial oxygen saturation (SaO_2), and systemic venous oxygen saturation ($SsvO_2$) in patients with a functionally single ventricle and flow of blood to the lungs provided by an aortopulmonary shunt.[1] The dashed line indicates the critical range for Q_p/Q_s, where systemic oxygen availability exceeds basal oxygen demands. These calculations are based on an assumed consumption of oxygen of 18 mL O_2/min, and an assumed weight of 3.0 kg. CO, cardiac output (right ventricular output in the case of a patient with hypoplasia of the left heart); $SpvO_2$, pulmonary venous oxygen saturation.

desaturation, due to hypoventilation, pulmonary oedema, or lung disease, results in a lower arterial oxygen saturation for a given ratio of pulmonary-to-systemic flow. Because low cardiac output and anaemia reduce the mixed venous saturation of oxygen, both of them influence arterial saturations of oxygen. Hence, it is not always possible directly to relate arterial saturations of oxygen to the ratio of pulmonary to systemic flow (Fig. 6.1),[1–3] although a high saturation generally implies high flow of blood to the lungs. There are a variety of manoeuvres available that influence the pulmonary and systemic vascular resistances.

Manipulation of Blood pH

Alkalosis, be it metabolic or respiratory, vasodilates the lungs, while constricting the systemic circulation, whereas the converse is also true, with acidosis being a pulmonary vasoconstrictor but a systemic vasodilator.[4,5] Blood pH can be manipulated via control of the tension of carbon dioxide in the blood, by adjusting the minute ventilation, or through infusion of base. The pH is kept in the range from 7.35 to 7.40 for patients with a tendency for increased pulmonary flow, usually by adjusting the ventilatory rate, and hence the arterial tension of carbon dioxide. For babies with excessive pulmonary vascular resistance, a range of pH from 7.5 to 7.6, achieved by reducing the arterial tension of carbon dioxide or by increasing the levels of bicarbonate in the serum, or both, may be helpful. Because hyperventilation increases the mean pressure in the airways, and thus may reduce cardiac output and cause barotrauma, levels of bicarbonate in the serum should be maintained at least in the normal range to reduce the need for supernormal rates of ventilation.

Manipulation of Alveolar Oxygen

Alveolar hypoxia is a powerful pulmonary vasoconstrictor, whereas hyperoxia appears to vasodilate the lung, at least under some circumstances.[6] Babies with excessive pulmonary flow, therefore, are ventilated with room air, or inspired gas with subambient inspired concentrations of oxygen between about 0.16 and 0.20, obtained by blending nitrogen with room air. Unfortunately, without elaborate monitoring it is difficult to know whether a hypoxic gas mixture reduces the flow of blood to the lungs, or merely reduces the pulmonary venous saturations of oxygen. We have found the use of hypoxic inspired gas only inconsistently helpful. Continuous monitoring of the inspired concentrations and arterial saturations of oxygen are very important when using hypoxic gas mixtures. For patients with insufficient arterial saturations, due to either high pulmonary vascular resistance or pulmonary venous desaturation, enhancing the inspired concentrations of oxygen is indicated.

In this respect, however, there are no intravenous pharmacological agents available that will selectively constrict the pulmonary, as opposed to the systemic, vascular bed. Dopamine and epinephrine, in clinically relevant doses, have little differential effect on these two vascular beds, or may even tend to constrict the systemic vascular bed more than the pulmonary.[7,8] Hence, these drugs may be helpful in the baby with hypoxemia secondary to insufficient systemic vascular resistance,or decreased cardiac output, but not for reducing excessive pulmonary flow.

Furthermore, there are no intravenous drugs that reliably and selectively vasodilate the pulmonary circulation, although agents such as tolazoline and prostaglandin I_2 have been sometimes useful in babies with pulmonary vasoconstriction. On the other hand, inhaled nitric oxide is a selective pulmonary vasodilator, and might be expected to increase the flow of blood to the lungs, and arterial saturations of oxygen, in at least some patients with increased pulmonary vascular resistance.[9] A trial of inhaled nitric oxide is appropriate when increased pulmonary vascular resistance causes significant hypoxemia in these patients. The test dose in 80 parts per million, but the dose should be decreased to the lowest effective level, and around 5 parts per million may be all that is required. Such low doses minimize the risk of pulmonary toxicity. Strict monitoring of inhaled levels of nitric oxide and nitrogen dioxide and levels of methemoglobin in the blood is necessary when using this therapy.[9]

Inspired carbon dioxide has also been described as improving the haemodynamic state of babies with hypoplasia of the left heart. The effect of inspired carbon dioxide on the pulmonary and systemic haemodynamic states in this setting, however, has been poorly characterised thus far, and this approach is of unproven efficacy. Our experience with this strategy has been limited, and we have found it to be of modest benefit, at best.

Inotropic and Vasoactive Drugs

When seen preoperatively, patients with hypoplasia of the left heart may require little or no inotropic support, especially if the diagnosis is made sufficiently early that cardiovascular collapse does not occur. There is right ventricular overload, nonetheless, since this ventricle supplies both systemic and frequently excessive pulmonary blood flow, and inotropic support not only may improve peripheral perfusion but also make fluid balance easier to maintain. If systemic blood pressure is sufficient, dobutamine at dose of around 5 mg/kg/min is a reasonable choice. Dopamine is more helpful in increasing blood pressure when it is low, but, as noted above, it may tend to increase flow of blood to the lungs when given in high doses. Epinephrine, at a dose of from 0.02 to 0.1 mg/kg/min is more effective for refractory hypotension. Digoxin may also serve to improve contractility in these hearts, but it has the disadvantage of a long half-life, difficulty in titration, and the potential for toxicity, especially in patients with renal dysfunction.

"Housekeeping" Issues

Because these patients have some degree of hypoxemia, and often somewhat reduced systemic perfusion, it is important to maintain an adequate hematocrit, between 35% and 50%, depending on the saturations of oxygen and the systemic perfusion. Diuretics may be helpful in maintaining fluid balance, especially in mechanically ventilated patients who have a tendency to accumulate fluid in the third space. Adequate nutrition is important. How this is most safely and effectively provided is unclear. Because of the potential for gut ischemia, due to retrograde ductal flow and reduced systemic perfusion, the risk for necrotizing enterocolitis may be increased in babies.[10] While this suggests that parental nutrition may be preferable in the preoperative infant with hypoplasia of the left heart, intravenous alimentation carries its own risks, such as sepsis, and the need for a central venous catheter. Feeding with breast milk may reduce the risk of infection.

When Shock Complicates the Initial Presentation

The principles noted above apply to those patients who present in shock, as well as to those who are well compensated at time of presentation, only more so because of the greater need for inotropic support, the need for sodium bicarbonate to correct metabolic acidosis, and so on. We delay operation in these infants until the organs have recovered their function, especially with regard to renal, hepatic, and cardiac systolic function.

Management of the Restrictive Atrial Septal Opening

There is a spectrum of infants with hypoplasia of the left heart having a restrictive atrial communication. Those with a highly restrictive atrial communication have very limited pulmonary blood flow, usually severe pulmonary oedema, and life-threatening hypoxemia, and require immediate intervention to relieve the restriction (Fig. 6.2). Emergency Norwood palliation can be undertaken, but

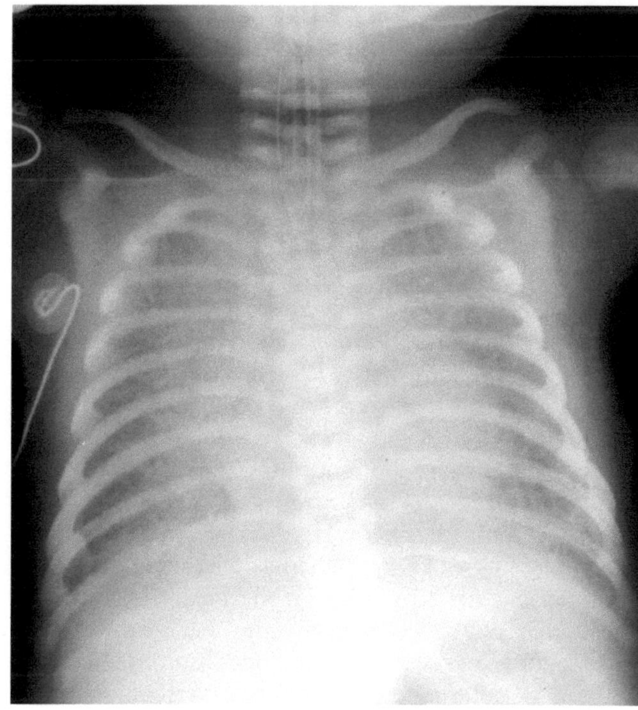

FIGURE 6.2. The chest radiograph from a patient with hypoplasia of the left heart and severe obstruction at the atrial septum. Severe pulmonary oedema is obvious.

because the pulmonary vascular resistance will be very high subsequent to bypass, and the lungs will be wet, severe hypoxemia following operation is likely. Therefore, this is a poor option. Enlarging the atrial septal communication in the catheterisation laboratory is potentially the most desirable approach, but it is technically difficult. The left atrium is often small, the interatrial septum thick, and the native interatrial opening quite superior and leftwardly deviated, all of which make standard Rashkind balloon septostomy relatively hazardous. It is possible, however, to enlarge the interatrial communication using a static balloon to enlarge the opening,[11] and we have successfully used this approach in several babies. A third option is to perform open atrial septostomy with the patient on the pump oxygenator, followed by a few days of recovery for pulmonary vascular resistance to fall and pulmonary oedema to resolve, with subsequent Norwood palliation. The approach chosen depends on the capabilities of the institution involved.

Patients with more modest restriction of the atrial septum, but adequate flow of blood to the lungs, do not require emergency intervention, but may be prone to significantly increased pulmonary vascular resistance immediately following Norwood palliation. While relatively prompt operation may be indicated for these patients, efforts to enlarge the interatrial communication would not appear to be warranted since, as noted above, these procedures carry a relatively high risk.

Selection of Patients

Are there cardiac or noncardiac complications that so reduce the likelihood of a satisfactory result that Norwood palliation may not be indicated? Are there patients who are clearly better treated with cardiac transplantation rather than Norwood palliation? We are unaware of any data that would allow definitive answer to these questions,[12] and can offer only our impressions. It might seem that patients who present in shock are at risk for sequels that may reduce the probability of good short-term results, for example, renal failure, or long-term outcome, for example, damage to the central nervous system. Even patients with a history of hypotension with severe metabolic acidosis, nonetheless, usually make good recoveries with appropriate resuscitation, and we have not excluded them from Norwood palliation, presuming return of the function of their organs. The size of the ascending aorta, while clearly a factor in determining the technical difficulty in achieving a good result, has not been significantly correlated with survival in our institution. We have not excluded from operation babies with right ventricular systolic dysfunction or tricuspid regurgitation, although these doubtless constitute significant risk factors. Patients with a severely obstructed atrial septum at the time of birth are at substantially increased risk for a poor long-term outcome, but if the obstruction can be promptly relieved, it may still be reasonable to offer Norwood palliation. While the likelihood of survival at our institution is significantly less for infants weighing less than 2.5 kg, successful operation even in babies less than 2.0 kg is possible. At what weight the chance of successful operation becomes prohibitive is unclear. We, and others, have found that infants with severe aortic obstruction, and dilated and poorly contractile left ventricles, do poorly after Norwood palliation, and generally have offered orthotopic transplantation of the heart to such patients.[13]

Timing of Operation

We have generally avoided operation in the first 1 or 2 days of life, thus permitting pulmonary vascular resistance to fall and thereby reducing the risk of severe postoperative hypoxemia, but we are aware of any data suggesting the earliest "optimal" time for Norwood palliation. Data from our institution suggests a higher risk of mortality in those undergoing surgery beyond 4 weeks of age, probably related to elevated pulmonary vascular resistance,[14] although increased lung water and other factors may also play a role. We anticipate that these older patients are likely to have marginal or inadequate arterial saturations postoperatively, and we are always prepared to execute the manoeuvres noted above. Consideration should also be given to construct a systemic-to-pulmonary shunt larger than would otherwise be used, although the risk of excessive flow of blood to the lungs, especially after the first several postoperative hours, may be significant.

POSTOPERATIVE MANAGEMENT

General Issues

In the early postoperative period, the requirements are the same as for any patient who has had surgery for a congenital cardiac defect, namely maintenance of adequate tissue perfusion and oxygenation, at a reasonable cost to the heart and lungs, while avoiding complications. Somewhat later, the emphasis shifts to timely extubation. How these requirements are achieved is also similar to how they are achieved in other postoperative patients, although there are some elements of postoperative care specific to the patient undergoing Norwood palliation.

Blood Pressure and Tissue Perfusion

At our institution, neonates who have had Norwood palliation have an average systolic arterial pressure of around 70 mm Hg, and diastolic pressure of around 35 mm Hg, for the first 24 hours postoperatively.[15] What should be considered a minimally acceptable arterial pressure for these babies is unclear. Neonates who have had complete cardiac repair, for example, the arterial switch operation for transposed great arteries, may tolerate arterial blood pressures as low as the systolic range of 50 mm Hg, but infants with the Norwood palliation are different. Because of the aortopulmonary shunt, flow of blood to the lungs is sensitive to systemic blood pressure, and may be insufficient at low arterial pressures. Coronary arterial perfusion pressure is probably also somewhat tenuous because of diastolic runoff into the lungs and a reduction in aortic diastolic pressure. For these reasons, it seems prudent to aim for arterial blood pressure in at least the range of 60 mm Hg systolic. On the other hand, excessive pressure imparts unnecessary ventricular work, and sometimes excess flow of blood to the lungs, implying that arterial pressures in excess of 80 mm Hg systolic may be higher than optimal. In fact, we have observed that sudden episodes of decreased systemic perfusion, acidosis, and inappropriately high arterial oxygen saturation may occur in the setting of chronically somewhat elevated systolic blood pressures, in the range from 80 to 90 mm Hg in a subset of these babies.

Arterial blood pressure may be acceptable, but perfusion to the organs may be suboptimal. There is no single measure of tissue perfusion for these or other patients with the colour and temperature of the skin, the amplitude of the peripheral pulses, the time to capillary refill, urinary output, and presence or absence of metabolic acidosis, taken together as a rough index of perfusion. Systemic blood flow can also be measured, although this is somewhat cumbersome since the conventional thermodilution technique cannot be used. There are certain caveats worth noting, however, when applying these measures of perfusion. Because these patients have diastolic flow into the aortopulmonary shunt, and hence a relatively wide pulse pressure, the pulses may seem full even with decreased systemic blood flow. Low urinary output sometimes occurs in patients with adequate systemic blood flow and pressure, which usually requires 24 to 48 hours for resolution, suggesting a transient fall in renal function, perhaps related to intraoperative events.

A variety of measures are available to manipulate blood pressure and tissue perfusion. It is important to provide adequate intravascular volume. Measurement of the central venous pressure is the single best measure of ventricular filling pressure in these patients, and needs to be adjusted taking the overall haemodynamic picture into account. For patients with adequate perfusion and pressure, a central venous pressure of around 4 to 5 mm Hg may be sufficient, while in patients with decreased ventricular function, levels of 10 to 12 mm Hg may be required. Because excess intravascular volume tends to find its way into the "third space" and the lungs, it is generally desirable to give no more fluid than is required for adequate cardiac output. The capacity to carry oxygen is also important, especially in patients with suboptimal tissue perfusion and/or arterial saturation. At least in the early postoperative period, a hematocrit of from 40% to 50% should be maintained.

Use of intravenous inotropic agents, as described above, is valuable. In our institution, patients following Norwood palliation are generally initially treated with from 5 to 10 μg/kg/min of dopamine, although dobutamine is occasionally used in the same dose range. Epinephrine given in low doses, from 0.02 to 0.1 μg/kg/min is often helpful in increasing arterial pressure in the baby who does not respond adequately to dopamine. We generally use milrinone as well, which does not increase blood pressure, but rather improves systemic perfusion.

Some practitioners routinely use phenoxybenzamine, an α-adrenergic antagonist, to decrease systemic vascular resistance in the postoperative period. Data from Tweddel and co-workers[16] suggests that this medication may have a favorable impact on systemic oxygen delivery, but there are insufficient data to establish whether or not this medication actually improves outcome in these patients. As noted below, how these patients are mechanically ventilated markedly affects blood pressure and perfusion.

Mechanical Ventilation

How the parameters for mechanical ventilation impact on the situation subsequent to the Norwood procedure depends to some extent on the characteristics of the reconstructed cardiac anatomy. We found a relatively high inspired concentration of oxygen, along with alkalosis, to be well tolerated in our postoperative neonates.[15] We attribute this to the fact that, in our patients, the flow of blood to the lungs seems to be limited by size and diameter, along with the origin and length, of the systemic vessel chosen for construction of the systemic-to-pulmonary arterial shunt. Pulmonary vascular resistance, therefore, is relatively unimportant in restricting pulmonary flow. In patients with functionally larger shunts, however, alveolar hyperoxia and alkalosis may be poorly tolerated (see below).

Delayed Sternal Closure

Leaving the sternum unapproximated immediately following surgery may improve the haemodynamic state of patients with significant chest wall and mediastinal oedema. The risk of infection of the surgical wound is is probably higher in patients with an open sternum.[17] Unpublished data from our institution have shown that aggressive diuresis, and thus timely sternal closure, is important. It is not necessary for these patients to be pharmacologically paralysed, although extubation must obviously await sternal approximation.

Mechanical Support of the Circulation

In postoperative Norwood patients with inadequate systemic perfusion due to transient reduction in myocardial function, support using extracorporeal life support, or a ventricular assist device, can be lifesaving. Extracorporeal life support can also be helpful for patients with an acutely thrombosed aortopulmonary shunt. At our institution, approximately two-fifths of the patients treated in this fashion subsequent to Norwood palliation survive to be discharged from hospital.

"Housekeeping" Issues

Analgesia

Fentanyl, at a dose of 6 μg/kg/h, is used for the first postoperative night, along with pharmacological paralysis with pancuronium, but this is changed to morphine sulphate the next day, starting at a dose of 30 μg/kg/h, and tapering off over the next 3 to 5 days if the patient is haemodynamically stable. Sedation is provided by lorazepam, sometimes alternating with chloral hydrate.

Nutrition

We generally start enteral feeding on the second or third postoperative day, assuming the patient has had no significant episodes of hypotension or hypoxia, and is doing well haemodynamically. Patients with recent or current cause for gut ischemic are given parental nutrition.

Antibiotic Prophylaxsis

For patients with a closed sternum, antibiotic prophylaxis begins with Kefzol administered 1 hour prior to the dermal incision, and continues until the chest tubes are removed.[18] Antibiotic prophylaxis for patients with an open chest is provided by vancomycin and gentamicin.

Other Medications

Diuretics are used to augment the flow of urine in the immediate postoperative period, and this is usually continued through hospital discharge. Because all patients undergoing Norwood palliation have a tendency toward, or even frank, congestive cardiac failure, digoxin is usually started prior to discharge from hospital. We have the impression that the inhibitor of angiotensin converting enzyme, captopril, may be useful in patients with congestive cardiac failure. This may be related to its effect on ventricular afterload, and perhaps also due to reducing flow of blood to the lungs secondary to its effect in decreasing systemic vascular resistance. Captopril is used also in patients who tend to have systolic blood pressures greater than about 80 mmm Hg.

Potential Anatomic Problems

Our experience has been that patients with technically adequate Norwood palliation usually have acceptable haemodynamics and are stable postoperatively. There should be a high index of suspicion, therefore, that babies with inadequate systemic blood flow or reduced arterial saturations of oxygen may have technically suboptimal palliation.

Residual Coarctation

Residual obstruction of the descending aorta, due to the surgical reconstruction being carried insufficiently distal, is an occasional postoperative finding, although with increasing experience with the Norwood palliation it is infrequent. Coarctation in this setting is particularly pernicious, as it not only increases right ventricular afterload, and reduces perfusion to the lower body, but also may cause excess flow into the aortopulmonary shunt, which is proximal to the coarctation.[19] These babies may have severe haemodynamic compromise and often require early reoperation. It is often difficult definitively to make this diagnosis in the infant with low systemic flow by physical examination or echocardiography, and angiography is usually necessary.

Inadequate Resection of the Interatrial Septum

If the interatrial septum is not widely excised, left atrial hypertension will limit the flow of blood to the lungs. As is the case with residual coarctation, this problem has been seen much less frequently as the need for wide excision of the atrial septum has been fully appreciated. It is also more likely to occur after a period of weeks or months, rather than in the early postoperative period. Nevertheless, in the baby with inadequate arterial saturations of oxygen, this potential problem needs to be ruled out, especially in those with pulmonary congestion. Echocardiography with Doppler is very useful in detecting significant obstruction.

Aortopulmonary Shunt Too Big

Excess flow to the lungs in the early postoperative period in these patients is poorly tolerated, and death from myocardial dysfunction can result within hours of surgery. There are multiple reasons for this. The right ventricle is subjected to a large volume load, which is particularly bothersome given that ventricular function is depressed following cardiopulmonary bypass. There is relative hypoxemia, so the excess of flow to the lungs "steals" from the systemic circulation, and reduces perfusion to multiple organs. Very importantly, myocardial perfusion is reduced due to decreased diastolic blood pressure and other factors. In fact, a study of myocardial perfusion in clinically well infants following Norwood palliation has shown them to have significantly less resting myocardial flow than infants of similar age with completely repaired heart lesions, and a reduced maximum coronary flow as assessed with adenosine.[20] This reduced myocardial perfusion is present despite increased systemic ventricular volume work in the patients with hypoplasia of the left heart. Thus, the myocardial reserves, both for contractility and coronary arterial perfusion, are limited in these infants, and there is relatively little margin for suboptimal physiology.

Patients with excessive flow of blood to the lungs have findings of inadequate tissue perfusion, especially metabolic acidosis and poor urinary output. While they generally have relatively high arterial saturations of oxygen, in excess of 80%, if the mixed venous saturation is very low because of low systemic blood flow or anaemia, the arterial saturation may not be inappropriately high.[1,3] The diastolic arterial pressure is usually low, less than 30 mm Hg. Although, as previously described, there are a variety of ventilatory and other manoeuvres that can minimize the problem of an excessively large shunt, if these prove insufficient, it may be necessary to carry out urgent surgical narrowing of the shunt.

Aortopulmonary Shunt Too Small

It is not uncommon for initial arterial saturations of oxygen to be relatively low, at around 60%, in the first few postoperative hours. This may be due to a number of factors, perhaps most importantly transiently increased pulmonary vascular resistance or abnormalities in ventilatory perfusion. As saturations generally increase over the first few hours, we are not usually concerned with initial values in the range of 60%, as long as other indexes of well-being are satisfactory, and there is a progressive increase in oxygenation. Persistently inadequate saturations, nonetheless, raise the possibility that the shunt is too small, as well as the considerations noted previously, either because the diameter of the graft is too small, or because the shunt has narrowed, almost always at one or another site of anastomosis. There can be inadequate flow through the shunt if the systemic connection is to a small vessel, for example, to a carotid artery in a patient with anomalous origin of the subclavian artery. These patients, who may have what seems to be a good shunt murmur, are largely unresponsive to increased inspired concentrations of oxygen and hyperventilation, may have diminished pulmonary vascular markings on the chest roentgenogram, and their arterial saturation of oxygen is quite sensitive to the arterial blood pressure. In this setting, the most effective means of increasing temporarily the saturation of oxygen over the short run is to maintain increased arterial pressure using dopamine or epinephrine, although this may not be well tolerated over the longer run. Echocardiography with careful color Doppler may be helpful in establishing whether the shunt is narrowed. Selective angiography is often needed, however, with injection of contrast into the shunt or the brachiocephalic artery, presuming this to be the artery of origin. Injection into the ascending aorta rarely adequately delineates the anatomy of the shunt (Fig. 6.3).

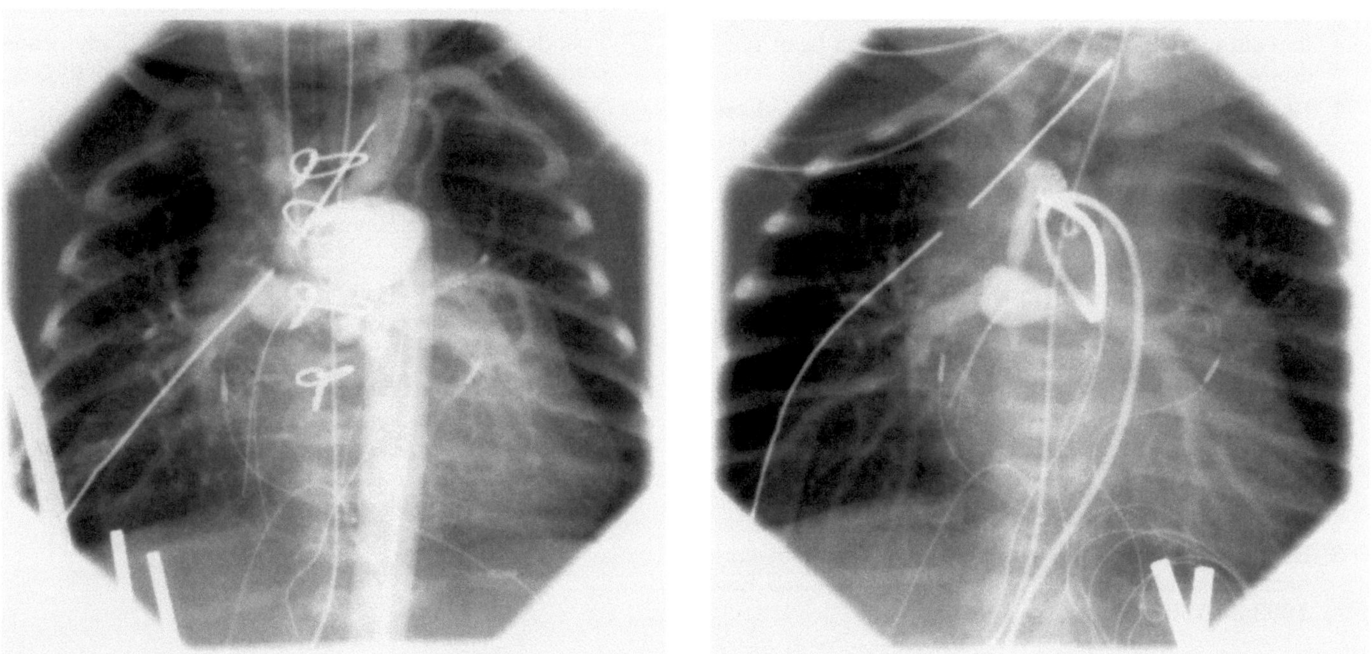

A B

FIGURE 6.3. Injection of contrast into the ascending aorta (A), and into the brachiocephalic artery (B), at the origin of a modified Blalock-Taussig shunt in a baby with unexpected hypoxemia following Norwood palliation. The injection into the aorta rarely outlines the usually subtle, but unmistakable, appearance of narrowing (arrow), and more selective injection is usually required.

Failure to Extubate Successfully

Any of the previously described structural lesions, by causing congestive heart failure, or an unacceptably low saturation of oxygen, may result in failure successfully to extubate. A neonate with a paralysed hemidiaphragm may do well on minimal ventilatory support, but require reintubation shortly after extubation because of an excess work of breathing. Plication of the paradoxically moving paralysed hemidiaphragm may considerably shorten the length of time needed for mechanical ventilation. Abnormalities of the large airways, or paralysed vocal chords, can also preclude extubation, but these have been unusual in our experience with babies subsequent to Norwood palliation of hypoplasia of the left heart. Suboptimal nutrition, especially in conjunction with one of the preceding, can also be important.

References

1. Barnea O, Austin EH, Richman B, Santamore WP. Balancing the circulation: theoretic optimisation of pulmonary/systemic flow ratio in hypoplastic left heart syndrome. J Am Coll Cardiol 1994;24:1376–1381.
2. Fahey JT, Lister G. Oxygen demand, delivery and consumption. In: Fuhrman BP, Zimmer JJ, eds, Paediatric Critical Care, St. Louis: Mosby Year Book, 1992:238.
3. Rossi AF, Sommer RJ, Lotvin A, et al. Usefulness of intermittent monitoring of mixed venous oxygen saturation after stage I palliation for hypoplastic left heart syndrome. Am J Cardiol 1994;73:1118–1123.
4. Gordon JB, Martinez FR, Keller PA, Tod ML, Madden JA. Differing effects of acute and prolonged alkalosis on hypoxic pulmonary vasoconstriction. Am Rev Respir Dis 1993;148:1651–1656.
5. Chang AC, Kucker HA, Hickey PR, Wessel DL. Pulmonary vascular resistance in infants after cardiac surgery: role of carbon dioxide and hydrogen ion. Crit Care Med 1995;23:568–574.
6. Custer JR, Hales CA. Influence of alveolar oxygen on pulmonary vasoconstriction in newborn lambs versus sheep. Am Rev Respir Dis 1985;132:326–331.
7. Lang P, Williams RG, Norwood WI, Castaneda AR. The haemodynamic effects of dopamine in infants after corrective cardiac surgery. J Paediatr 1980;96:630–634.
8. Barrington K, Chan W. The circulatory effects of epinephrine infusion in the anaethetised piglet. Pediatr Res 1993;33:190–193.
9. Kulik TJ. Inhaled nitric oxide in the management of congenital heart disease. Curr Opinion Cardiol 1996; 11:75–80.
10. McElhinney DB, Hedrick HL, Bush DM, et al. Necrotizing enterocolitis in neonates with congenital heart disease: risk factors and outcomes. Pediatrics 2000;106:1080–1087.
11. Atz AM, Feinstein JA, Jonas RA, Perry SB, Wessel DL. Preoperative management of pulmonary venous hypertension in hypoplastic left heart syndrome with restrictive atrial septal defect. Am J Cardiol 1999; 15:1224–1228.
12. Jenkins PC, Flanagan MF, Sargent JD, et al. A comparison of treatment strategies for hypoplastic left heart syndrome using decision analysis. J Am Coll Cardiol 2001;38:1181–1187.
13. Lloyd TR. Prognosis of the hypoplastic left heart syndrome. Prog Paediatr Cardiol 1996;5:57–64.
14. Iannettoni MD, Bove EL, Mosca RS, et al. Improving results with first stage palliation for hypoplastic left heart syndrome. J Thorac Cardiovasc Surg 1994; 107:934–940.
15. Mosca RS, Bove EL, Crowley DC, Sandhu SK, Schork MA, Kulik TJ. Haemodynamic characteristics of neonates following first stage palliation for Hypoplastic left heart syndrome. Circulation 1995; 92(S 11):11-267–11-271.
16. Tweddel JS, Horrman GM, Fedderly RT, et al. Phenoxybenzamine improves systemic oxygen delivery after Norwood procedure. Ann Thorac Surg 1999; 67:161–168.
17. Pollock E, Ford-Jones EL, Rebeyka I, et al. Early nosocomial infections in paediatric cardiovascular surgery patients. Crit Care Med 1990;18:378–384.
18. Maher KO, VanDerEltzen K, Bove EL, Mosca RS, Chenoweth CE, Kulik TJ. A retrospective review of three antibiotic prophylaxis regimens for pediatric cardiac surgical patients. Ann Thorac Surg 2002; 74:1195–200.
19. Jonas RA, Lang P, Hansen D, Hickey P, Castaneda AR. First stage palliation of hypoplastic left heart syndrome. J Thorac Cardiovasc Surg 1986;92:6–13.
20. Donnelly JP, Raffel DM, Shulkin BL, et al. Resting coronary flow and coronary flow reserve in human infants following repair or palliation of congenital heart defects as measured by positron emission tomography. J Thorac Cardiovasc Surg 1998;115:103–110.

Chapter 7

SURGICAL MANAGEMENT OF HYPOPLASIA OF THE LEFT HEART

Edward L. Bove

Once considered a uniformly fatal condition, refinements in operative technique and perioperative care have been largely responsible for an improved outlook for newborns with hypoplasia of the left heart. Staged reconstruction for such patients is based on the knowledge that an effective circulation is possible in the absence of a subpulmonary ventricle. The principle that the flow of blood to the lungs could be maintained when certain well-defined haemodynamic criteria were met has led to the application of the Fontan procedure to virtually all forms of functionally univentricular hearts.[1] The first stage in the reconstructive process, the Norwood procedure, must provide unobstructed pulmonary venous return across the atrial septum, unobstructed flow of blood into the systemic circuit from the morphologically right ventricle, and sufficient flow of blood to the lungs in the absence of significant volume overload.[2] This leaves the pulmonary and systemic circulations in parallel, with the morphologically right ventricle performing the increased work of providing volume for both circulations. The second-stage procedure, the hemi-Fontan or bidirectional Glenn operation, results in removal of the volume overload on the right ventricle imposed by the systemic-to-pulmonary arterial shunt, and the connection of the superior caval vein to the undivided pulmonary arteries.[3,4] Augmenting the central pulmonary arteries, avoiding disturbances of conduction, and constructing a potential connection for the inferior caval vein to the pulmonary arteries are essential components of the hemi-Fontan procedure. The second stage reconstructive procedure is also the optimal time for correction of additional risk factors, including tricuspid valvar regurgitation, and maintains a stable circulation free of right ventricular volume overload and pulmonary hypertension. During the last stage of reconstruction, namely completion of the Fontan procedure, inferior caval venous return is channelled to the pulmonary arteries to complete the separation of the pulmonary and systemic circulations. Although the current techniques have resulted in substantial improvements in the quantity and quality of survival, efforts to refine each stage of the process continue to evolve with increasing follow-up and evaluation.

THE NORWOOD PROCEDURE

The indications for the Norwood operation include not only the classic form of hypoplasia of the left heart characterised by aortic and mitral atresia, but also those variants with varying degrees of underdevelopment of the left-sided structures in which survival with the postoperative establishment of biventricular circulations is judged impossible or highly unlikely. These morphologic subgroups include hearts with hypoplasia of the aortic and/or mitral valves, hypoplasia of the left ventricular cavity, and hearts with common atrioventricular junction and atrioventricular septal defect with left ventricular hypoplasia. Also included are those patients with discordant ventriculoarterial connections in the setting of univentricular atrioventricular connection to a dominant left ventricle, where subaortic obstruction at the ventricular septal defect is either already present or is likely to develop. This is especially likely when there is coexisting obstruction in the aortic arch. In our institution, the management of this latter group of patients consists of a primary Norwood procedure rather than banding of the pulmonary trunk combined with repair of the aortic arch. For the purposes of this discussion, however, I will consider only those patients with the classic form of hypoplasia of the left heart, namely those with a right ventricular dependent circulation in addition to atresia or hypoplasia of the aortic valve.

Although the majority of newborns with hypoplasia of the left heart are well palliated with an infusion of prostaglandin and reduction of the inspired concentrations of oxygen to avoid high systemic arterial saturations, this condition cannot be sustained indefinitely. A progressive deterioration in the clinical state, characterised by increasing pulmonary edema, decreased systemic blood flow, and development of pulmonary vascular obstructive disease, will become evident with time. This will occur sooner in the presence of significant obstruction to pulmonary venous return, most commonly resulting from a restrictive atrial septal defect. In practice, when the systemic arterial oxygen saturation is in excess of 80%, peripheral perfusion is well maintained, and the chest radiograph shows little or no pulmonary oedema, these patients usually remain stable, and may undergo operation electively within a few days. More significant degrees of obstruction to pulmonary venous return, however, will result in arterial saturations below 75% to 80%, with severe and rapidly progressive pulmonary oedema. More urgent intervention is required before the pulmonary status deteriorates further. Satisfactory atrial septostomy can often be achieved in the catheterisation laboratory, often with the use of a biopsy forceps to open an atrial defect, and has frequently been successful in our laboratory. It has been our experience that optimal results following the Norwood procedure are obtained when the flow of blood to the lungs is restricted by constructing a relatively small shunt.[5] Satisfactory postoperative oxygen saturations are increasingly difficult to maintain, however, when a small shunt is used in a patient with pulmonary oedema. A larger shunt will be required, which will then often result in pulmonary overcirculation after only a few hours following cardiopulmonary bypass, concomitant with resolution of the pulmonary oedema and falling pulmonary vascular resistance. A rapid deterioration in systemic perfusion may result in hypotension, metabolic acidosis, and death. We showed

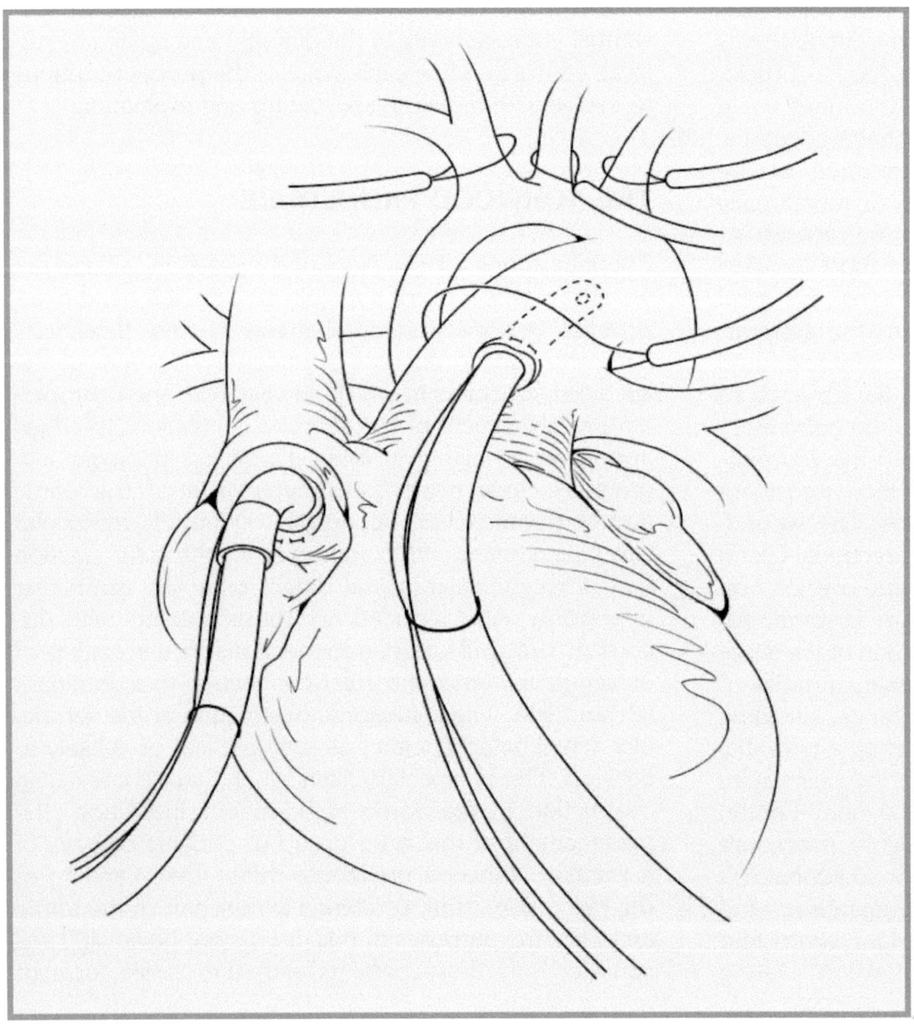

FIGURE 7.1. The arterial cannula is inserted into the arterial duct, and the atrial cannula through the right atrial appendage. The duct is snared with a tourniquet to exclude the pulmonary circulation as cardiopulmonary bypass is begun. Tourniquets are also shown around the arch vessels.

in an earlier study that early survival following first-stage palliation fell dramatically when patients underwent operation beyond the first month of life.[5]

Surgical Technique

Although the method of surgical repair has remained essentially constant over the past few years, a number of modifications have continually improved the quality of the outcome[5–9]:

- accurate orientation of the alignment between the proximal ascending aorta and the pulmonary trunk,
- augmentation of the descending aorta well beyond the distal insertion of the arterial duct,
- improved tailoring of the pulmonary allograft patch used to augment the aortic arch,
- the use of smaller systemic-to-pulmonary arterial shunts, and
- the recent introduction of shunts placed directly from the right ventricle to the pulmonary arteries.

Cardiopulmonary bypass is established by direct cannulation of the arterial duct and right atrial appendage (Fig. 7.1). A snare is tightened around the duct and the arterial cannula to exclude the pulmonary circulation, and systemic cooling is begun in preparation for circulatory arrest or regional cerebral blood flow. The ascending, transverse, and proximal descending segments of the aorta are mobilised, and a cryopreserved pulmonary allograft is trimmed to fashion a patch that will serve to augment the aorta and allow an anastomosis to the proximal pulmonary trunk. An allograft should be selected, if possible, in which the right and left pulmonary arteries are at least 1 to 2 cm long so that the contour matches the inner curvature of the aortic arch. Because pulmonary allograft material dilates when exposed to systemic pressures, the patch should not be cut too wide, or else the ascending aorta may dilate excessively, thus compressing the left pulmonary artery. The pulmonary trunk is then divided distally, and the bifurcation is closed with a patch to prevent narrowing of the right and left pulmonary arteries (Fig. 7.2). If regional cerebral perfusion is used, the proximal anas-

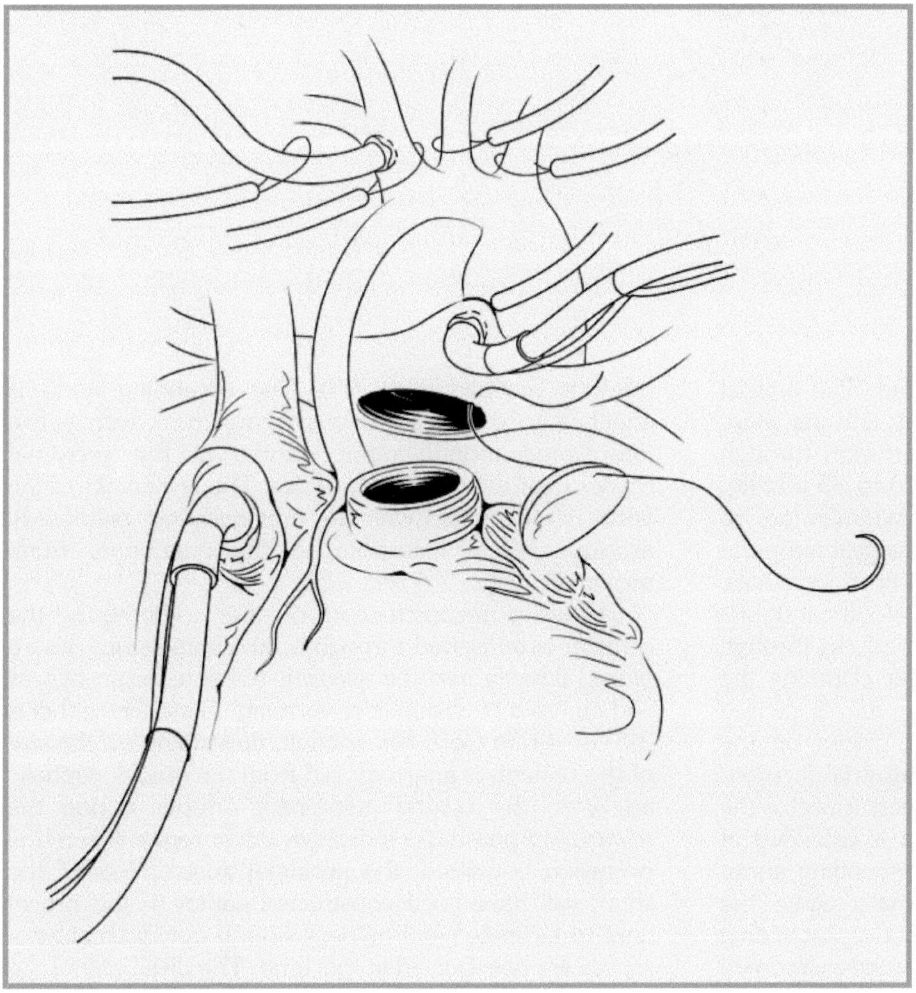

FIGURE 7.2. While cooling on cardiopulmonary bypass, the main pulmonary trunk is divided and the distal end closed with a patch. Shown in this illustration is the completed proximal anastomosis of the shunt, which can be used as access for regional cerebral perfusion (see text for details). After the initiation of circulatory arrest or regional cerebral perfusion, the atrial septum is excised (not shown).

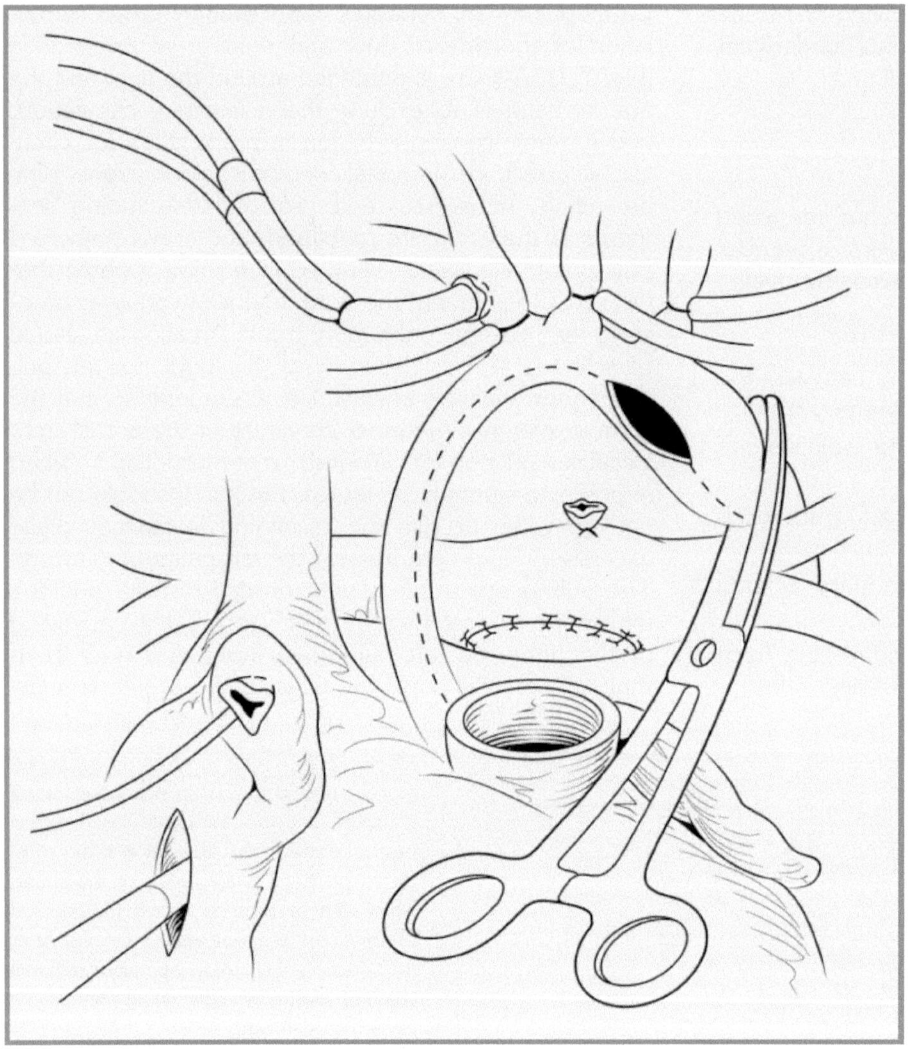

FIGURE 7.3. The duct is ligated proximally, and all additional ductal tissue is excised from the undersurface of the aortic arch. This opening is extended distally for approximately 10 to 15 mm, and proximally to the level of the divided pulmonary trunk. Illustrated here is regional cerebral perfusion with a cardiotomy suction catheter in the right atrium.

tomosis of the shunt is now constructed. The arterial cannula is then advanced into the shunt, and the snare adjusted to allow selective cerebral perfusion through the brachiocephalic artery at flows of 20 to 25 mL/kg. If not, the circulation is arrested after a minimum period of cooling of 20 minutes at a nasopharyngeal temperature of less than 18° to 20°C, and the brachiocephalic vessels are occluded with snares. Cold blood cardioplegia is administered at a total dose of 40 mL/kg through a side arm on the arterial cannula after clamping the descending thoracic aorta. The atrial septum is excised either through the purse-string suture placed on the atrial appendage or through a separate atrial incision. All ductal tissue is excised from the undersurface of the aortic arch, and the resultant opening is extended at least 10 mm further distally into the descending aorta. This incision is then extended proximally under the transverse arch and down the diminutive ascending aorta until the level of the previously divided pulmonary trunk is reached (Fig. 7.3). The ascending aorta is attached to the adjacent pulmonary trunk with a few interrupted monofilament sutures in the proximal corner near the coronary arteries. The remainder of the aorta is augmented with the previously cut pulmonary allograft patch, incorporating the pulmonary trunk proximally (Figs. 7.4 and 7.5).

Following reconstruction of the aortic arch, the cannula is reinserted through a new purse-string suture placed directly into the allograft tissue to begin bypass and commence systemic rewarming. Generally, either a 3.5- or 4-mm Gore-Tex conduit, depending on the size of the patient, is anastomosed from the brachiocephalic artery to the central pulmonary arteries during the rewarming phase. As indicated, when regional cerebral perfusion is utilized, the proximal anastomosis of the shunt will have been constructed earlier in the procedure to facilitate cerebral perfusion. If not, both anastomoses are constructed at this time. The distal end of the

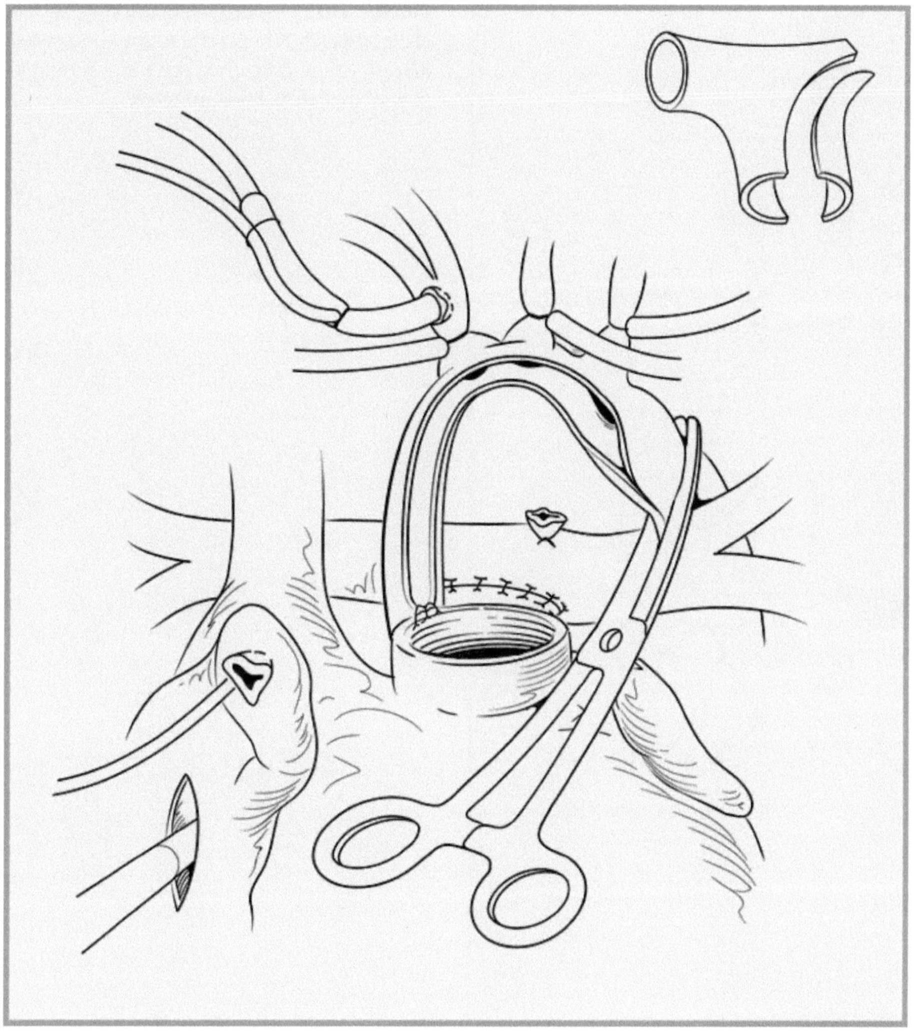

FIGURE 7.4. Augmentation of the aortic arch is accomplished with a patch of tissue cut from a pulmonary allograft. The proximal corner of the aorta is attached to the adjacent pulmonary artery with a few interrupted sutures. Care must be taken here to avoid any distortion or obstruction to coronary blood flow.

shunt is placed centrally, close to the divided end of the duct, rather than onto the right pulmonary artery itself (Fig. 7.6).

Recently, preliminary reports have indicated that a shunt constructed between the right ventricle and the bifurcation of the pulmonary trunk is associated with superior haemodynamic performance, since this eliminates the diastolic run-off inherent in a shunt placed between a systemic artery and the pulmonary arteries. Early survival was increased in some[10] but not all[11] reports using this approach. Presumably, the higher diastolic pressure associated with shunts placed from the right ventricle to the pulmonary arteries leads to improved coronary arterial flow and better right ventricular function.

Weaning from cardiopulmonary bypass is usually accomplished with infusions of intravenous dopamine, milrinone, and epinephrine. In the presence of a systemic-to-pulmonary arterial shunt, it is important to reduce afterload so as to lower systemic vascular resistance relative to pulmonary vascular resistance and decrease excessive flow of blood to the lungs. Optimal systemic oxygen saturation when bypass is discontinued is between 75% and 80%, with tension of oxygen from 28 to 35 mm Hg. These data can be properly evaluated, however, only when mixed venous saturation is also known, to estimate the difference in arterial and venous saturations of oxygen. When systemic arterial oxygen saturation is less than 65%, it is important to determine if the cause is elevated pulmonary vascular resistance or a technical problem with the shunt itself. The former will improve with time, while the latter will require operative revision. Poor right ventricular function may be the result of inadequate coronary arterial perfusion, and may also require operative revision. Delayed sternal closure is employed, with elective sternal closure performed in the intensive care unit a few days after surgery.

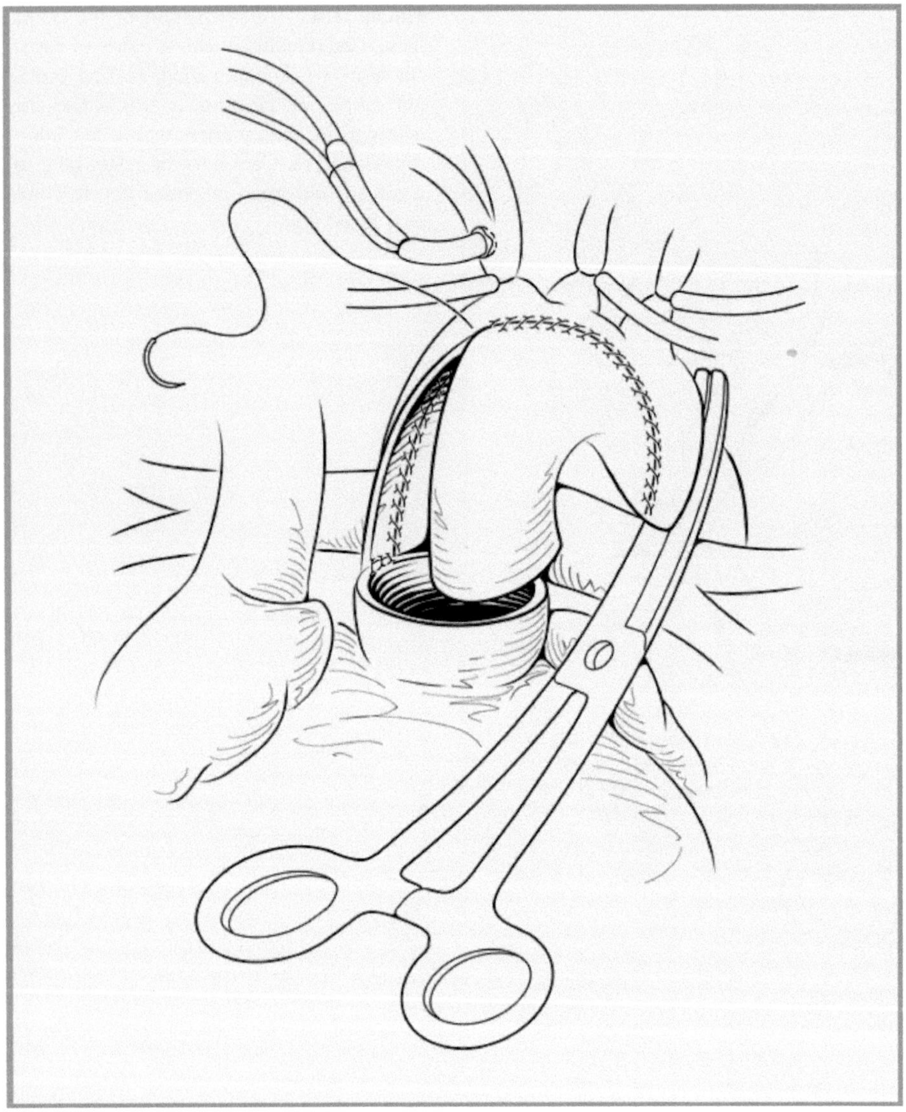

FIGURE 7.5. Appropriate tailoring of the allograft patch is important to avoid excessive dilation of the ascending aorta and compression of the left pulmonary artery.

THE HEMI-FONTAN PROCEDURE

At the University of Michigan, patients surviving the first stage of palliation undergo a hemi-Fontan procedure between 4 and 6 months of age.[3] The use of small shunts at the initial operation has occasionally necessitated the need to proceed to the second stage at an even earlier age, as the patients "outgrow" their pulmonary blood flow. We have performed the hemi-Fontan as early as 1 month of age, preferring this alternative to the use of larger initial shunts or the addition of a second systemic-to-pulmonary arterial shunt after the first few postoperative months. In a review of all hemi-Fontan procedures performed in patients under 6 months of age between December 1990 and February 1995 at the University of Michigan, survival was 94%, specifically 81 of 86 patients, and was not different for those patients with hypoplasia of the left heart compared to other diagnoses.[12] Of our patients, 14 were under 3 months of age, with 13 of these surviving. Younger age was a risk factor for hypoxemia occurring within the first 48 hours after operation, and for pulmonary arterial thrombosis, which occurred in two patients, but not for death.

Cardiac catheterisation is routinely performed to assess:

- right ventricular function,
- tricuspid regurgitation,
- residual obstruction of the aortic arch or pulmonary venous flow across the atrial septum,
- pulmonary arterial anatomy, and
- pulmonary vascular resistance.

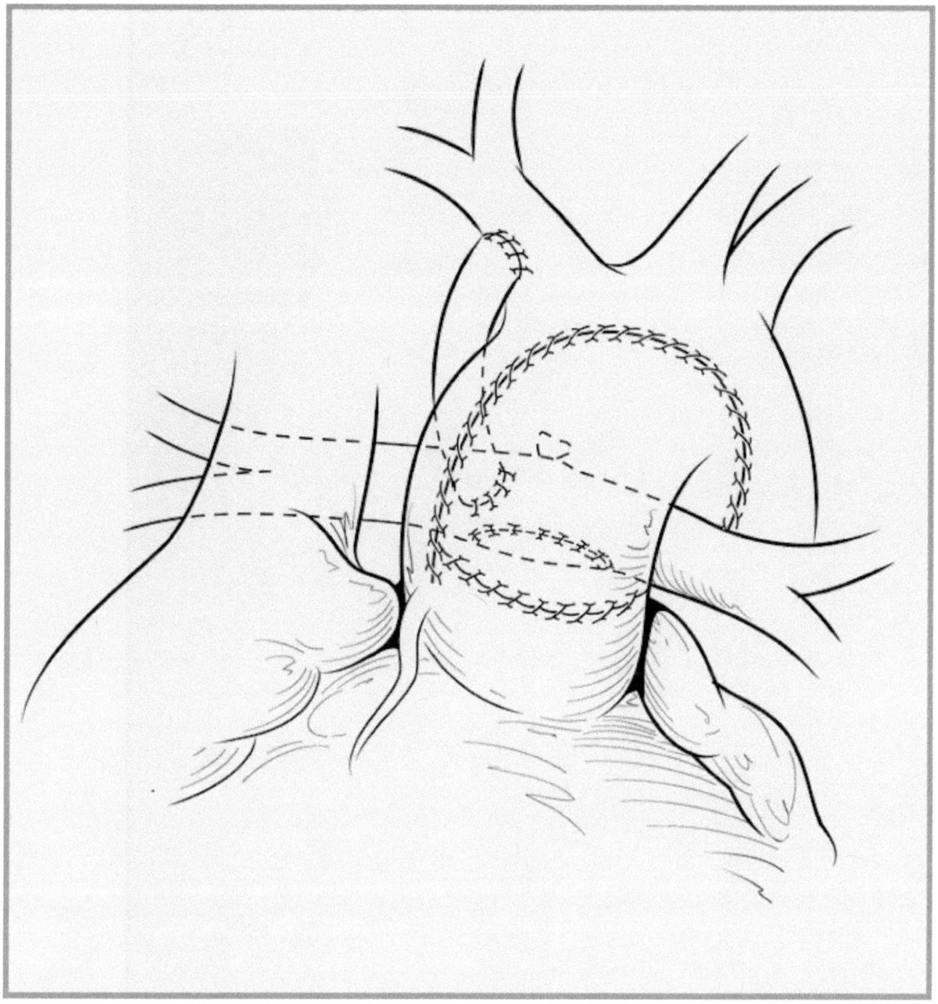

Figure 7.6. The final reconstruction demonstrates the position of the shunt, which is placed between the brachiocephalic artery and the central pulmonary arteries.

The number of patients with contraindications to proceeding with the second-stage procedure has been surprisingly few. Absolute contraindications have included isolated poor right ventricular function and elevated pulmonary vascular resistance. When right ventricular function has been poor in association with another correctable problem, typically residual obstruction of the aortic arch, excessive volume overload from a large shunt, or tricuspid valvar regurgitation, we have proceeded with the hemi-Fontan with the expectation that ventricular function will improve. The expectation that tricuspid valvar regurgitation was secondary to volume overload, and that it would improve after the second-stage procedure, has not proven to be the case. Thus, we now perform tricuspid valvoplasty at the hemi-Fontan procedure when regurgitation is more than mild.[13] Although placement of the shunt on the central pulmonary arteries near the ductal insertion has resulted in their more symmetric growth following the Norwood procedure, it has been common to find mild narrowing at the origin of the left pulmonary artery (Fig. 7.7). The use of pulmonary allograft patches of smaller widths has also improved the reconstruction of the aortic arch, and reduced the incidence of compression of the left pulmonary artery (Fig. 7.8).

Surgical Technique

Cannulation for cardiopulmonary bypass is accomplished through the distal ascending aorta, placing the arterial cannula through allograft patch tissue, and the venous cannula through the right atrial wall, lowering the perfusate temperature to 25°C. A small wire-reinforced vent is placed in the superior caval vein at its junction with the brachiocephalic to prevent venous hypertension and assist the venous drainage for later in the procedure when the right atrial cannula is removed. Alternatively, two separate cannulas may be placed directly in the caval veins. The shunt is divided, and the pulmonary arteries are mobilised widely. A patch of pulmonary allograft material is fashioned for augmentation of the pulmonary arteries, similar in shape to the

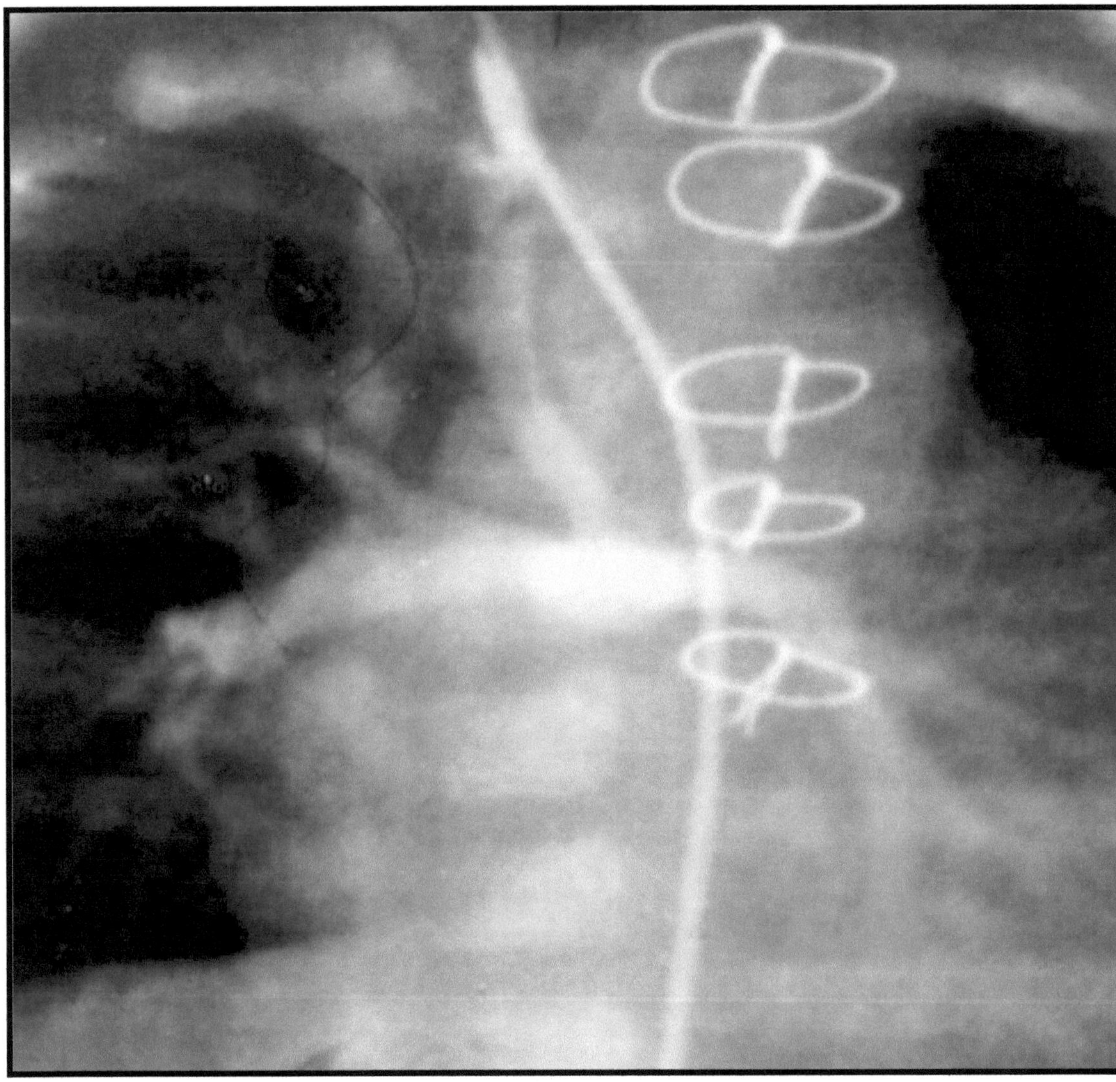

Figure 7.7. This angiogram, taken 6 months after first-stage reconstruction, demonstrates the optimal position of the shunt, permitting symmetric growth of the pulmonary arteries.

patch used in the Norwood procedure. The superior caval vein is exposed, and the azygous vein is ligated. The central pulmonary arteries are opened, beginning beneath the superior caval vein and extending nearly to the branch to the left upper lobe (Fig. 7.9). The right atrial cannula is clamped, and an incision is made in the base of the right atrial appendage, but is stopped before its junction with the superior caval vein. During this period, bypass is maintained with venous return through the vent in the caval vein, and hand-held suckers inserted into the right atrium itself. This incision has been modified to avoid injuring the artery to the sinus node. A patch is placed within the right atrium, at the level of the superior rim of the oval fossa, which will isolate superior caval venous return into the pulmonary arteries, and provide an unobstructed pathway for the connection of inferior caval venous return at the time of the Fontan procedure (Fig. 7.10). Prior to the insertion of this patch, the atrial septal defect is inspected, and enlarged if necessary. This is best accomplished by cutting back the coronary sinus into the left atrium. If indicated, the tricuspid valve is repaired at this time. The cannula providing venous return is reinserted into the atrium, and systemic rewarming is begun. The pulmonary artery is sutured to the outside of the superior caval vein until the right atrium is reached (Figs. 7.10 and 7.11). The pulmonary arteries are then enlarged with the allograft patch.

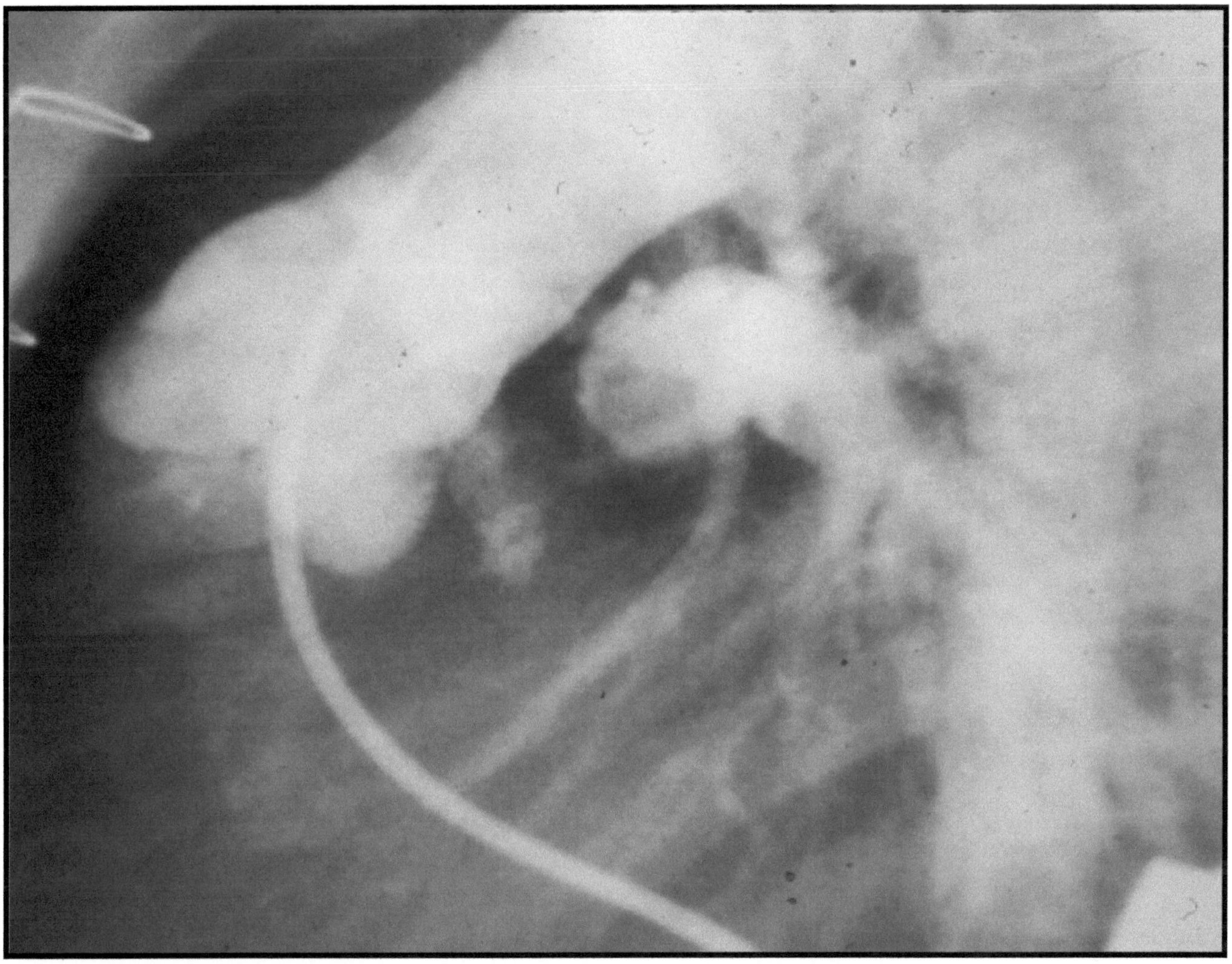

Figure 7.8. This angiogram is performed with contrast injected into the reconstructed ascending aorta, and is shown in the lateral view. The normal contour of the aortic arch is maintained without distal obstruction or excessive dilation. The native proximal aorta and the pulmonary arteries can be seen posterior to the ascending aorta. (From Bove and Mosca,[9] with permission.)

THE FONTAN PROCEDURE

Following the second-stage reconstruction, the majority of patients have maintained satisfactory palliation for a considerable period of time.[14] Some, however, will experience a progressive decline in systemic arterial oxygen saturation with the development of venovenous collaterals to the inferior caval venous circulation. Additionally, increasing systemic hypoxemia is noted in some patients as a result of increasing growth and activity, leading to an increased return of desaturated blood from the lower body with exercise. Those patients with azygous continuation of an interrupted inferior caval vein have also experienced a progressive fall in oxygen saturation as venous collaterals have formed to the hepatic venous circulation. Arteriovenous malformations may constitute another important etiology for progressive hypoxemia in this group.[15] For all these reasons, the Fontan procedure has generally been performed at approximately 2 years of age. Preoperative evaluation, including Doppler, echocardiography, and cardiac catheterisation, is essential to assess ventricular function, atrioventricular valvar regurgitation, pulmonary arterial size, pressure, and resistance, and to check on the presence of systemic-to-pulmonary collateral arteries. Major collateral arteries should be embolised preoperatively. The size of the left pulmonary artery is of potential concern following staged reconstruction for these patients (Fig. 7.12). A dilated neoaorta following the Norwood procedure may cause compression and subsequent hypoplasia of the left pulmonary artery. In addition, the situation of low pressure

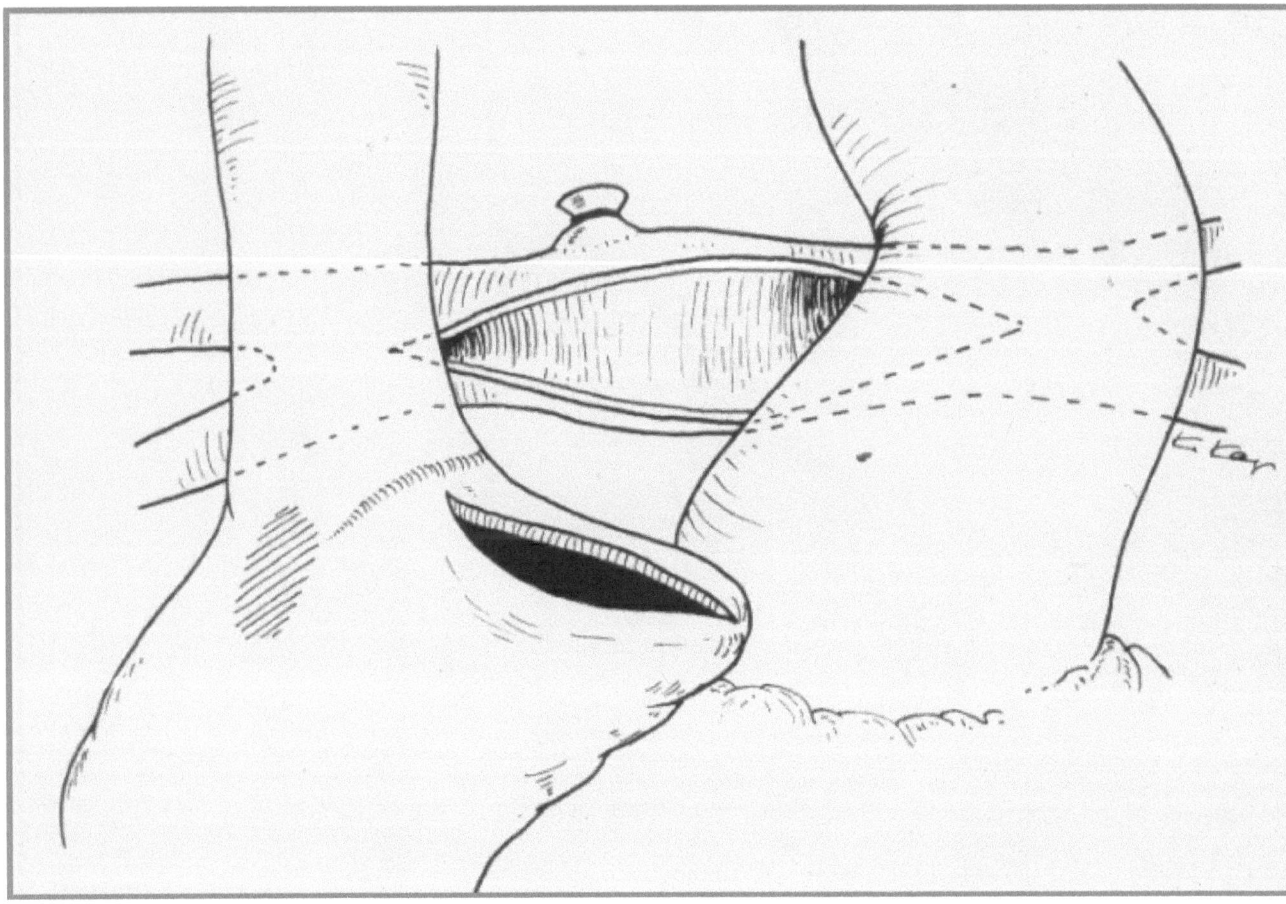

FIGURE 7.9. The initial incisions for the second-stage reconstruction. The central pulmonary arteries are opened anteriorly, and an incision is made along the base of the right atrial appendage without crossing the cavoatrial junction to avoid injuring the artery to the sinus node.

and flow imposed by the hemi-Fontan procedure may result in the left pulmonary artery being of insufficient size for the Fontan circulation. Insertion of an endoluminal stent at the time of operation has proved to be a useful adjunct in a few of our patients with otherwise refractory hypoplasia of the left pulmonary artery. We prefer to utilise stents sparingly, and, when necessary, to place them at the final procedure, because a larger stent can then be safely inserted.

Surgical Technique

The heart is again exposed through a midline sternotomy, dissecting adhesions sparingly to avoid injury to the phrenic nerves. Venous return is accomplished with cannulation directly through the inferior caval vein, and the procedure is done during full-flow bypass at a perfusate temperature of 28°C. A vent is placed into the pulmonary arteries through the augmenting patch placed previously at the time of the hemi-Fontan procedure. Following placement of the aortic cross clamp, and the administration of cardioplegia, the right atrium is opened and the internal patch previously placed between the superior caval vein, the right atrium, and the pulmonary arteries is removed.[16] The sizes of the right and left pulmonary arteries are checked if necessary. Dilation with calibrated metal dilators, or insertion of stents, may be performed at this time if the left pulmonary artery is restrictive. A Gore-Tex patch is then fashioned to channel inferior caval venous return to the pulmonary arteries (Figs. 7.13 and 7.14). Fenestration has been performed in our institution for all patients considered at high risk for the Fontan operation. This is accomplished by placing a 4-mm defect in the Gore-Tex baffle. Spontaneous closure occurs frequently, although it may also be performed in the catheterisation laboratory.

RESULTS

Previous reports have analysed operative experience with first-stage palliation for classic hypoplasia of the left

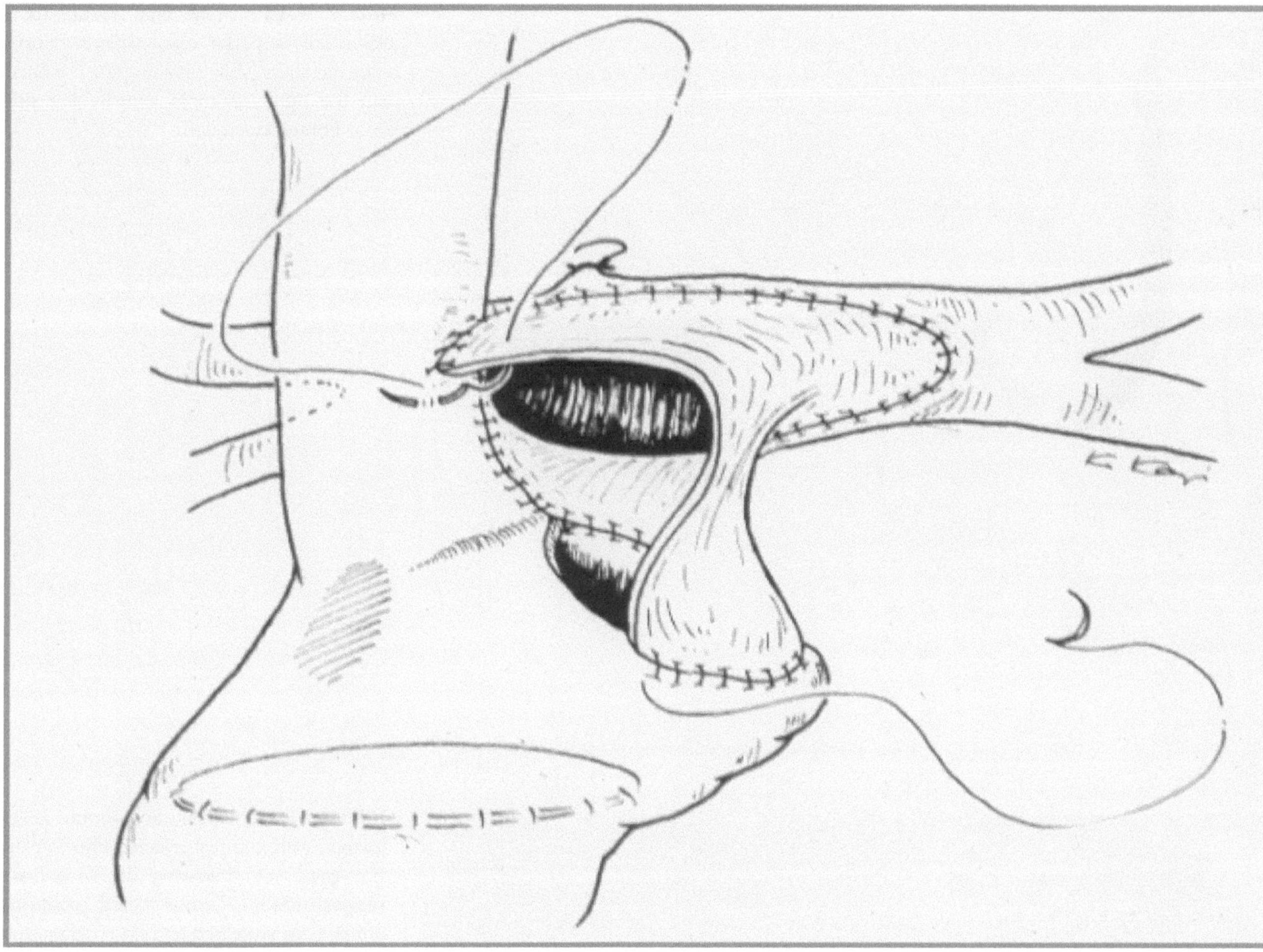

FIGURE 7.10. A patch is fashioned from a cryopreserved pulmonary allograft and used to augment the central pulmonary arteries. An additional patch is placed within the right atrium to exclude the superior caval vein and the pulmonary arteries from the heart.

heart at our institution.[3,5,6,12] In these reports, earlier date of operation and older age at operation were identified as risk factors for early death. With an increasing number of patients, additional risk factors have emerged and include

- severe obstruction to pulmonary venous return,
- low birth weight or prematurity, and
- the presence of significant noncardiac abnormalities.

Anatomical subtypes and the diameter of the ascending aorta have not been risk factors.[5] The experience with patients undergoing staged reconstruction for the classic form of left heart hypoplasia between 1990 and 1995 was previously reported.[17] Among the 158 patients undergoing surgery during this time frame, 127 were considered at standard risk, having no significant associated cardiac or noncardiac condition, absence of severe obstruction to pulmonary venous return, and age less than 1 month. In this group, hospital survival following the Norwood procedure was 86%, with the 70% confidence limits between 82% and 89%. During the time frame of this study, there were 103 hospital survivors among 106 patients undergoing the hemi-Fontan procedure, giving a survival of 97%, with the 70% confidence limits between 95% and 98%. The Fontan procedure was performed in 80 patients, with 70 survivors, giving survival of 88%, and the 70% confidence limits between 83% and 91%. Actuarial survival through all operative procedures, including intervening deaths unrelated to operation, was 69%, with standard deviation of 8%, at 5 years of age for patients with typical anatomy, and 71% with a deviation of 17% for those with variant anatomy.

As a means of comparison to more contemporary outcomes for the surgical palliation of hypoplasia of the left heart at the University of Michigan, we have reviewed the results for the past 100 consecutive

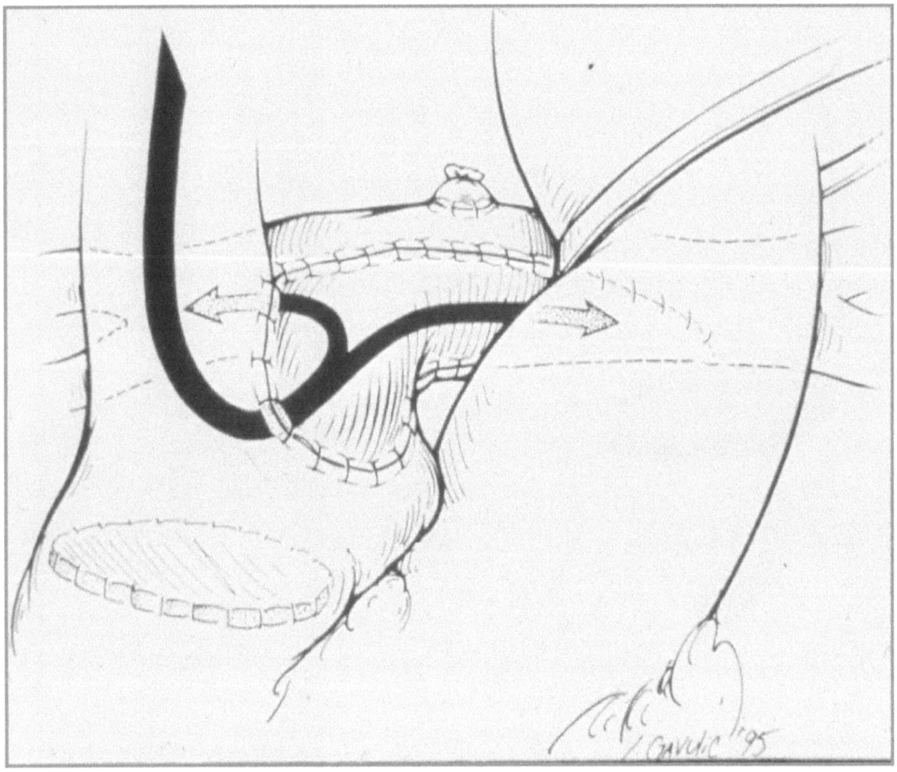

FIGURE 7.11. The final reconstruction allows only superior caval venous return to enter the lungs while retaining the connection with the right atrium to simplify the subsequent Fontan operation.

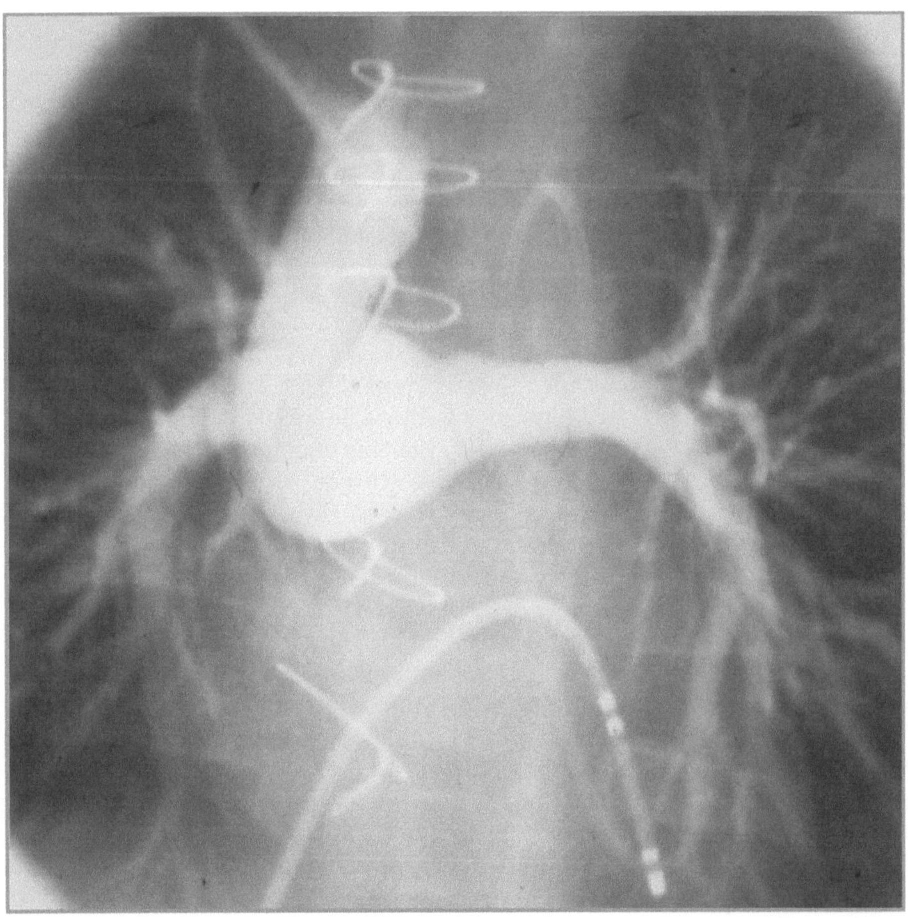

FIGURE 7.12. This angiogram is performed with contrast injected into the superior caval vein following the second-stage reconstruction. Unobstructed pulmonary arteries are visualised and no contrast enters the right atrium. (From Bove and Mosca,[9] with permission.)

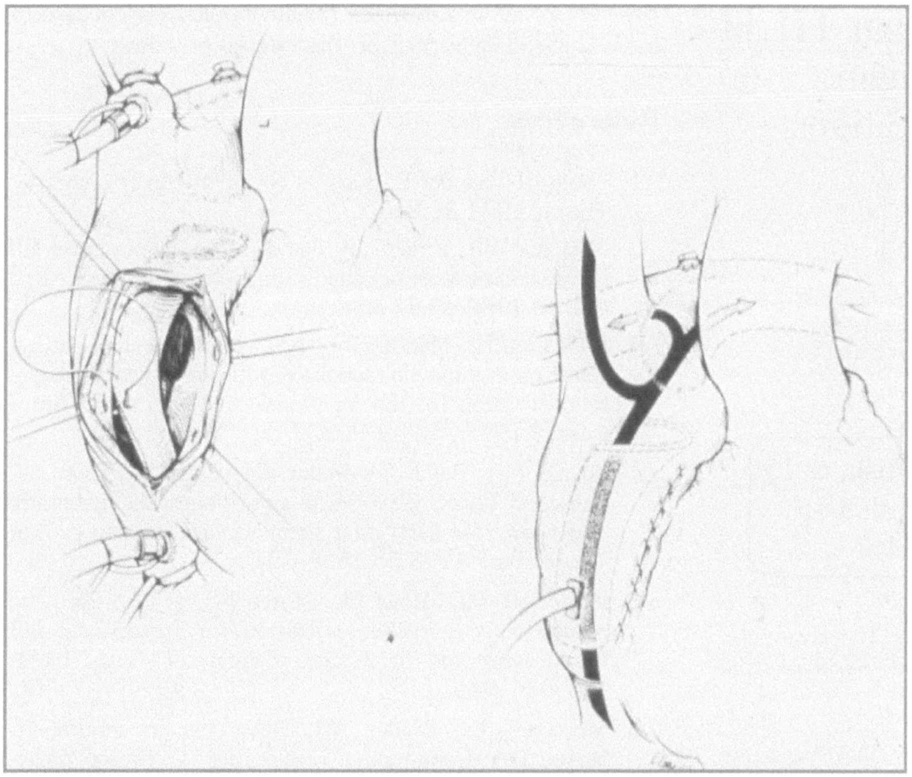

FIGURE 7.13. The Fontan procedure is performed by first removing the previously inserted patch beneath the connection between the superior caval vein and the pulmonary arteries. The coronary sinus is cut back to enlarge the atrial septal defect if necessary. A polytetrafluoroethylene patch is cut to the appropriate length and width to fashion a patch that will channel inferior caval venous return to the previously constructed anastomosis with the pulmonary arteries. (From Bove and Mosca,[9] with permission.)

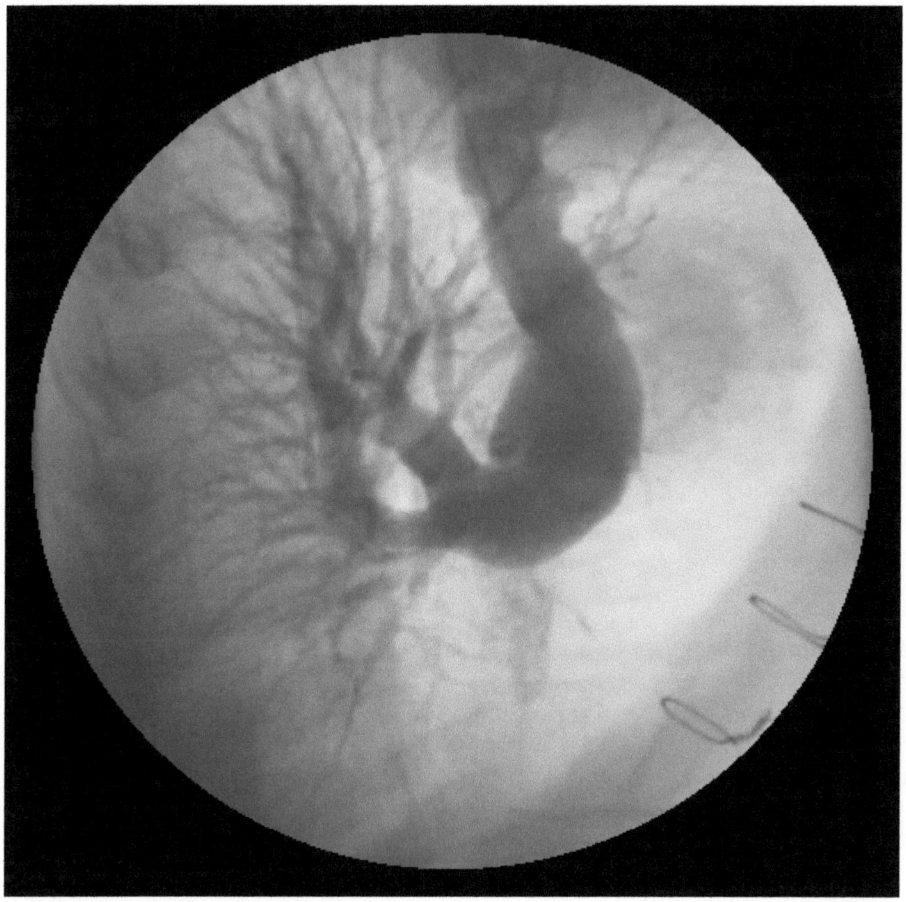

FIGURE 7.14. A lateral angiogram performed 6 months after the Fontan procedure demonstrating the completed cavopulmonary connection.

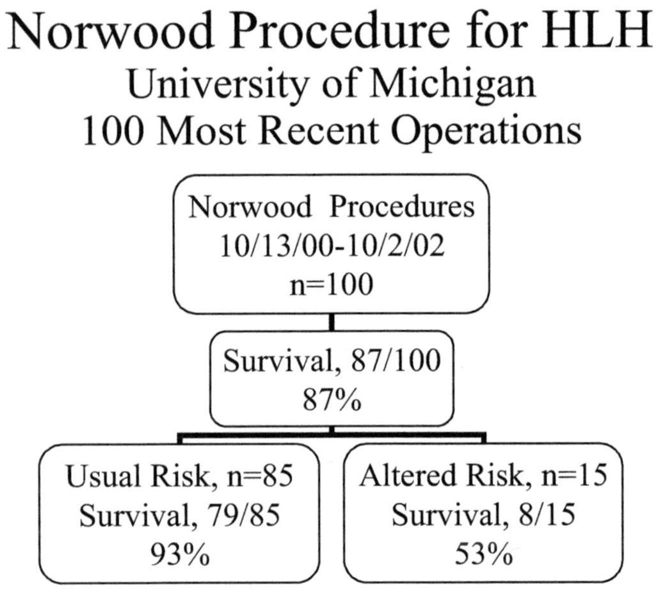

FIGURE 7.15. The results of the most recent 100 consecutive Norwood procedures for hypoplastic left heart (HLH) syndrome at the University of Michigan.

Norwood procedures for the classic phenotype (Fig. 7.15). These patients underwent surgery between October 2000 and October 2002. Hospital survival was 87% for the entire group, with 93% of those with standard risk surviving, specifically 79 of 85, compared to 53% survival of those considered to be at greater risk, specifically 8 of 15. The results for the Fontan procedure have also improved, with only one patient dying from a postoperative neurological event in the last 100 consecutive procedures.

A recent modification in the technique of constructing the systemic-to-pulmonary shunt appears promising as a means of reducing mortality for certain groups considered to be at high risk. The shunt is placed directly between the right ventricle and the pulmonary arteries, avoiding the immutable haemodynamic disadvantage of the diastolic runoff that occurs with any systemic-to-pulmonary arterial shunt. The higher diastolic blood pressure is likely to result in improved coronary arterial flow and better right ventricular function. Preliminary reports have noted a greater degree of haemodynamic stability with this shunt, obviating the need to manipulate systemic and pulmonary vascular resistance to balance the relative flows between the two circulations.[10,11] The possible consequences, however, of an incision placed in the morphologically right ventricle in patients with hypoplasia of the left heart must also be considered. Because the early mortality following the first-stage reconstruction remains the greatest period of risk, improvement in these results would be expected to have significant impact on the overall outcome.

References

1. Fontan F, Baudet E. Surgical repair of tricuspid atresia. Thorax 1971;26:240.
2. Norwood WI, Kirklin JK, Sanders SP. Hypoplastic left heart syndrome: experience with palliative surgery. Am J Cardiol 1980;45:87–91.
3. Pridjian AK, Mendelsohn AM, Lupinetti FM, et al. Usefulness of the bidirectional Glenn procedure as staged reconstruction for the functional single ventricle. Am J Cardiol 1993;71:959–962.
4. Seliem MA, Baffa JM, Vetter JM, Chen SL, Chin AJ, Norwood WI Jr. Changes in right ventricular geometry and heart rate early after hemi-Fontan procedure. Ann Thorac Surg 1993;55:1508–1512.
5. Iannettoni MD, Bove EL, Mosca RS, et al. Improving results with first-stage palliation for hypoplastic left heart syndrome. J Thorac Cardiovasc Surg 1994; 107:934–940.
6. Meliones JN, Snider AR, Bove EL, Rosenthal A, Rosen DA. Longitudinal results after first-stage palliation for hypoplastic left heart syndrome. Circulation 1990;82(suppl IV):IV-151–IV-156.
7. Jonas RA, Lang P, Hansen D, Hickey P, Castaneda AR. First-stage palliation of hypoplastic left heart syndrome. J Thorac Cardiovasc Surg 1986;92:6–13.
8. Pigott JD, Murphy JD, Barber G, Norwood WI. Palliative reconstructive surgery for hypoplastic left heart syndrome. Ann Thorac Surg 1988;45:122–128.
9. Bove EL, Mosca RS. Surgical repair of the hypoplastic left heart syndrome. Prog Paediatr Cardiol 1996;5:23–35.
10. Pizarro C, Malec E, Maher KO, Januszewska K, Gidding SS, Norwood WI. Right ventricle to pulmonary artery conduit improves outcome after Norwood procedure for hypoplastic left heart syndrome. Circulation 2002;1069(9):II–393.
11. Sai H, Kado H. Alternative procedure of modified Norwood operation for hypoplastic left heart syndrome. Circulation 2002;1069(19):II–694.
12. Bradley SM, Mosca RS, Hennein HA, Crowley DC, Kulik TJ, Bove EL. Bidirectional superior cavopulmonary connection in young infants. Circulation, 1996;94(suppl II):II-5-II-11.
13. Reyes A, Bove EL, Mosca RS, Kulik TJ, Ludomirsky A. Tricuspid valve repair in children with hypoplastic left heart syndrome during stage reconstruction. Circulation 1996;94:I–297.
14. Chang AC, Farrell PE Jr, Murdison KA, et al. Hypoplastic left heart syndrome: hemodynamic and angiographic assessment after initial reconstructive surgery and rele-

vance to modified Fontan procedure. J Am Coll Cardiol 1991;17:1143–1149.

15. Srivastava D, Preminger T, Lock JE, et al. Hepatic venous blood and the development of pulmonary arteriovenous malformations in congenital heart disease. Circulation 1995;92:1217–1222.

16. Norwood WI Jr, Jacobs ML, Murphy JD. Fontan procedure for hypoplastic left heart syndrome. Ann Thorac Surg 1992;54:1025–1030.

17. Bove EL, Lloyd TR. Staged reconstruction for hypoplastic left heart syndrome: contemporary results. Ann Surg 1996;224(3):387–395.

Chapter 8

EVOLVING TECHNIQUES IN THE OPERATIVE MANAGEMENT OF HYPOPLASIA OF THE LEFT HEART

Akhlaque N. Bhat, Aitizaz U. Syed, and Marco Pozzi

The principles of the physiological repair of patients having hypoplasia of the left heart have essentially remained the same since Norwood published his first report two decades ago.[1] From a procedure that was performed in a few centres, it is now performed in most of the major paediatric cardiac centres in the world. Naturally, the procedure has evolved, both in surgical techniques as well as in the management of cardiopulmonary bypass. The improved outcomes currently seen after the Norwood procedure have resulted to some extent from the way the operation is conducted, together with the marked improvements in perioperative care. Nearly two decades of experience with the Norwood procedure and an equal duration of follow-up available from the earliest survivors of this operation have gone a long way toward clarifying the complex physiological balance that needs to be achieved for a favourable outcome. This chapter discusses the details of some of the trends that aim at modifying the classic Norwood procedure in ways that might lead to an improved outcome.

SURGICAL TECHNIQUES

The Norwood operation for the first stage of palliation of hypoplasia of the left heart is a generic procedure with three important goals:

- Construction of an unobstructed outflow from the systemic ventricle.
- Ensuring an adequate and unobstructed coronary arterial circulation.
- Construction of a systemic-to-pulmonary arterial shunt to establish a parallel pulmonary circulation, and an unrestrictive interatrial communication for unobstructed pulmonary venous return.

Though the goals of operation remain the same for all surgeons, the techniques by which they may be achieved vary.

Construction of an Unobstructed Outflow Tract from the Systemic Ventricle

In the classical Norwood operation, construction of an unobstructed outflow tract from the systemic ventricle involves anastomosing the diminutive aorta to the pulmonary trunk, filleting open the lesser curve of the small aortic arch from the ascending aorta right down to the descending aorta a few millimetres beyond the coarctation, and then augmenting the arch with homograft material to construct a neoaorta. Homograft material is generally regarded as being devoid of the potential to grow, hence raising the fear of the occurrence of late obstruction within the aortic arch. Some surgeons, therefore, have suggested modifications that eliminate the use of homograft tissue in the reconstruction of the aortic arch.

In 1995, Brawn's group[2] in the United Kingdom, and later that year Mee's group[3] in the United States, described a technique for primary reconstruction of the arch. Both groups have since published intermediate term results of this technique. The technique involves generous mobilisation of the ascending aorta, arch, descending aorta, and brachiocephalic arteries. The pulmonary trunk is transected obliquely, proximal to the origins of the right and left pulmonary arteries. The distal part of the trunk is closed with autologous pericardium. The proximal end of the modified

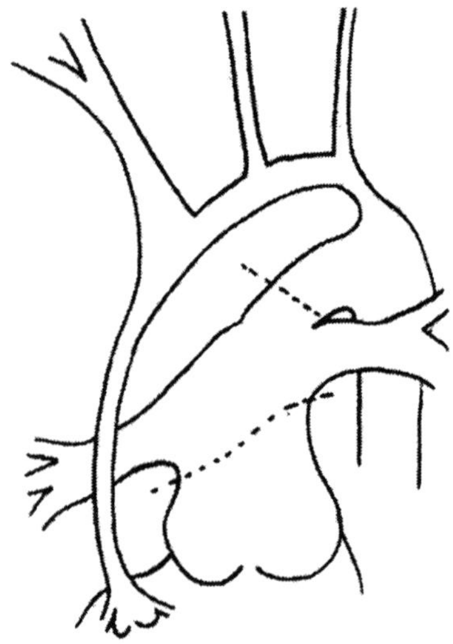

FIGURE 8.1. The modified Norwood procedure. The ascending aorta, arch, descending aorta, and the brachiocephalic arteries were completely mobilised. The pulmonary artery trunk was transacted obliquely, proximal to the right and left coronary arteries. (From Poirier et al.,[8] with permission.)

Blalock-Taussig shunt is fashioned on the distal brachiocephalic artery during cooling, which is instituted at 18°C. The heart is arrested with a single dose of cardioplegia. All ductal tissue is resected, and the lesser curvature of the arch is opened from the left subclavian artery to the ascending aorta. The arch is reconstructed by anastomosis of the descending aorta to the aortic isthmus and distal arch. The proximal pulmonary trunk is then anastomosed to a combination of ascending aorta, proximal arch, and descending aorta (Figs. 8.1 and 8.2). A tension-free reconstruction is possible by extensively mobilising the descending aorta, arch, and brachiocephalic arteries. The longitudinal split of the medial descending aorta and the use of as much of the length of the pulmonary trunk as possible are technical features that further reduce tension on the anastomosis.

Critics of this technique cite compression of the left pulmonary artery beneath the reconstructed neoaortic arch as the main objection to its use. There are also anecdotal reports about the compression of the main stem of the left bronchus under the neoaortic arch, resulting in obstructive emphysema and a need for revision of the repair. Although Mee's group[3] does not mention how many of its patients developed compression of the left pulmonary artery, the group does emphasise the need to evaluate the size of this artery by angiography before constructing the bidirectional cavo pulmonary shunt, and of confirming the finding during the operation. If the artery is found to be compressed, it needs to be augmented with an autologous pericardial patch, or else stented.

How important is it to reconstruct the neoaortic arch without using homograft tissue? Are the risks of long-term arch obstruction after the use of homograft tissue real? Mahle and associates[4] studied echocardiograhically the characteristics of growth of the reconstructed aortic

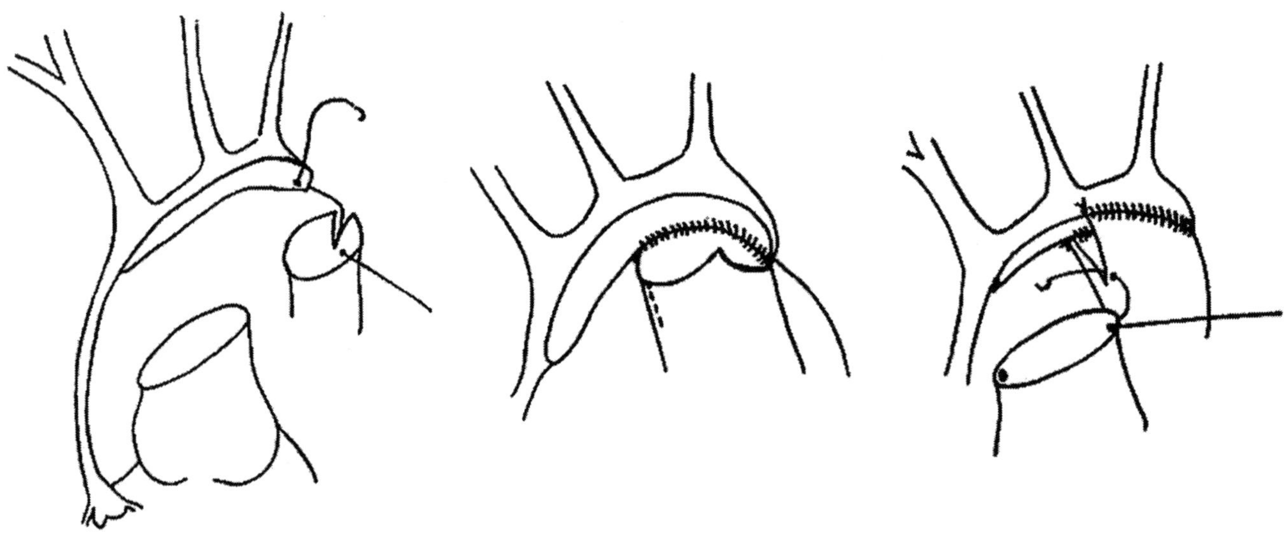

A B

FIGURE 8.2. The modified Norwood procedure (continued). A: Once all of the ductal tissue was resected, the lesser curvature of the arch was opened from the left subclavian artery to the ascending aorta. B: The arch was reconstructed by anastomosis of the descending aorta to the aortic isthmus and distal arch. The proximal pulmonary trunk was then sutured to a combination of ascending aorta, proximal arch, and descending aorta. (From Poirier et al.,[8] with permission.)

arch 10 years after the Norwood procedure. Their studies demonstrate that growth of the reconstructed aorta parallels growth seen in normal subjects. Their postmortem studies reveal that all of the growth of the reconstructed aorta occurs in the native tissue that makes up a part of the circumference of the aorta throughout its length. Fraisse et al.,[5] while studying the accuracy of echocardiography for detection of obstruction of the aortic arch after a Norwood procedure, found this complication in one-fifth of their 139 patients. In all, homograft material had been used to reconstruct the aortic arch. They concluded that the likelihood of obstruction tends to plateau beyond the first six postoperative months. Ishino et al.[6] found obstruction in almost one-quarter of their 95 patients in whom the arch had been reconstructed exclusively with the use of autologous tissue. Assuming meticulous surgical technique was employed, it can be argued that obstruction of the aortic arch after the Norwood procedure may more likely to be a result of factors other than regression of residual ductal tissue or the use of homografts or autologous tissue. Longer follow-up studies will certainly help shed light on, if not completely resolve, these contentious issues.

Ensuring an Adequate and Unobstructed Coronary Arterial Circulation

In the Norwood procedure, the native ascending aorta functions as a coronary artery. In those patients with aortic and mitral atresia, who account for more than half of those with hypoplasia of the left heart, the native ascending aorta is diminutive. It is vitally important, therefore, that the technique of its anastomosis to the neoaorta be such that the flow of blood to the coronary arteries is unobstructed. In a postmortem review of 122 patients who died after the Norwood procedure, Bartrum et al.[7] found impairment of coronary perfusion as the most frequent cause of death, being seen in one-quarter of their cohort. Precoronary stenosis was present in almost four-fifths of these patients, indicating a fault with the surgical technique that was used to connect the native ascending aorta to the neoaorta. In the classic Norwood procedure, the adjacent walls of the native ascending aorta and the pulmonary trunk are incised vertically and sutured to each other to keep widely patent the approach to the coronary arterial orifices. These investigators recommended that, at the completion of the aortopulmonary anastomosis, the pathway to the arterial orifices should remain patulous, and that the surgeon should be able to see the atretic aortic root and the adjacent arterial openings. With the diminutive aortas typically seen in patients with aortic and mitral atresia, this is not always possible.

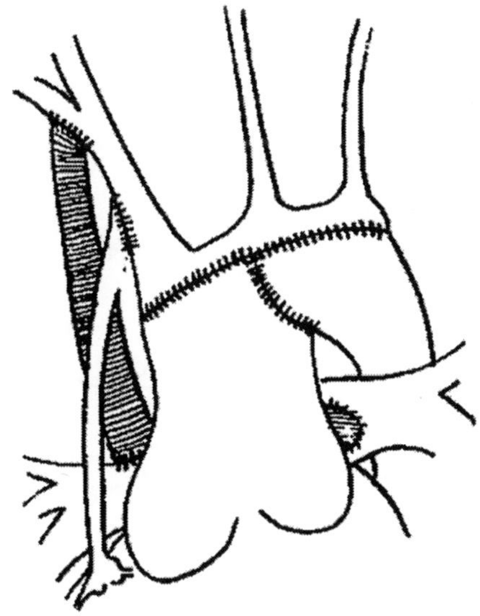

FIGURE 8.3. The modified Norwood procedure—the first modification. The ascending aorta was transected, shortened, and reimplanted onto the proximal brachiocephalic artery. (From Poirier et al.,[8] with permission.)

Consequently, modifications have emerged that have tried to address this problem. Mee's group,[8] in its sustained attempts to develop an effective technique, went through two modifications before settling on a third. In the first modification, the ascending aorta was transected, shortened, and reimplanted into the proximal brachiocephalic artery (Fig. 8.3). Of the seven patients in whom this technique was used, three died postoperatively. The authors attributed the deaths to a possible coronary steal from the adjacent modified Blalock-Taussig shunt. In the second modification, the ascending aorta was shortened and implanted onto the posterior face of the pulmonary trunk, preserving the lie of the ascending aorta (Fig. 8.4). Early deaths occurred in two of the three patients on whom this technique was used, and it was quickly abandoned. Compression of the native ascending aorta leading to myocardial ischaemia was believed to be the cause of the deaths. The third modification, which is the technique currently used by the group, consists of implantation of the full-length native ascending aorta onto the rightward and anterior sinus of the pulmonary trunk (Fig. 8.5). Of a total of 13 patients who had operations with this modification, two died.

Mechanical obstruction to coronary arterial flow is not the only cause of myocardial ischaemia in patients who have undergone the Norwood procedure. In the postmortem study of Bartram et al.,[7] four-fifths of the patients who had evidence of mechanical impediment to coronary arterial flow also had histological signs of

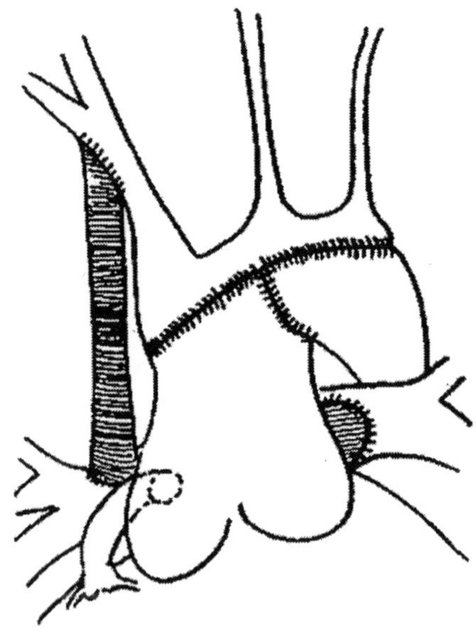

FIGURE 8.4. The modified Norwood procedure—the second modification. The ascending aorta was shortened and implanted onto the posterior face of the pulmonary trunk, preserving the lie of the ascending aorta. (From Poirier et al.,[8] with permission.)

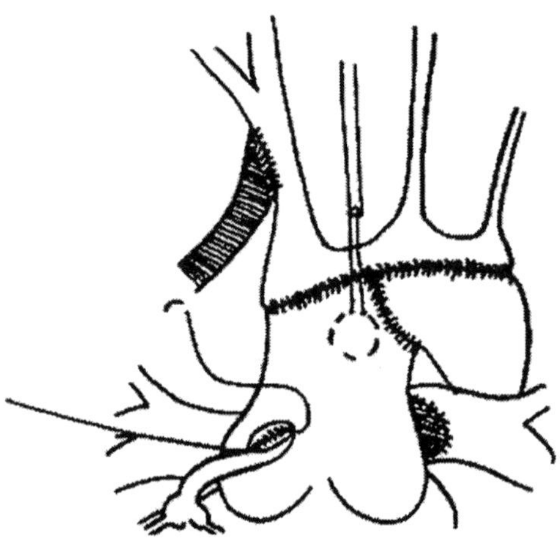

FIGURE 8.5. The modified Norwood procedure—the third modification. The ascending aorta was anastomosed onto the anterior rightward sinus of the pulmonary trunk. (From Poirier et al.,[8] with permission.)

myocardial ischaemia. A significant number of patients, nonetheless, had histological evidence of myocardial ischaemia without any evidence of mechanical obstruction to their coronary arterial flow, incriminating mechanisms other than obstruction as the cause of the myocardial ischaemia. The presence of a shunt-dependent pulmonary circulation is a major weakness in the Norwood procedure because, by permitting a diastolic runoff, this keeps the pulmonary and coronary circulations in constant competition. In such a scenario, it would be reasonable to assume that, while the presence of a systemic-to-pulmonary arterial shunt sets the stage for coronary steal, uncompromising patency of the coronary circulation assumes paramount importance.

Construction of the Systemic-to-Pulmonary Arterial Shunt

The means of establishing the flow of blood to the lungs has perhaps been the part of the Norwood procedure that has undergone the most evolution. Balancing pulmonary and systemic flows is critical to success. During the evolution of this procedure, shunts as varied as a 12-mm Hancock conduit, 8-mm nonvalved conduits, and 7- and 5-mm Gore-Tex shunts have been used. It is not difficult to imagine that a majority of these patients had problems resulting from excessive pulmonary blood flow. With a better understanding of post-Norwood physiology, the competitive nature of the flow through the shunt with that in the coronary and systemic circulations was established as the major cause of bad outcomes. This led surgeons to use smaller shunts, with a 3.5-mm Gore-Tex shunt considered adequate for a neonate weighting from 3 to 3.5 kg. The size of the shunt, however, is not the sole determinant of flow of blood to the lungs in these patients. While the size of the subclavian or the brachiocephalic artery and presence of an aberrant right subclavian artery determine the dynamics of flow through the shunt, preoperative factors like the condition of the lungs, pulmonary vascular resistance and any coexisting disease can also influence its size and selection.[9] Ventilatory adjustments, and pharmacological manipulation, are routinely used, with mixed results, to balance the flows in the systemic and pulmonary circulations of patients subsequent to the Norwood procedure. Selecting the correct size, however, is the cornerstone for achieving such a balance. Since the natural course of pulmonary vascular resistance in neonates allows for excessive blood flows as the baby grows older, it is not surprising that, in a majority of cases, surgeons find themselves downsizing shunts in the immediate postoperative period.

Schmid et al.[10] used an adjustable snare to balance the flow through the shunt. Of a total of 26 patients undergoing the Norwood procedure, seven had an adjustable tourniquet placed around the systemic-to-pulmonary arterial shunt. All seven patients survived. Of these seven patients, two had the tourniquet adjusted more than once in the postoperative period as a bedside procedure in the intensive care unit. Although in five patients the snare could be removed during delayed sternal closure, in the two patients who required readjustments the snare remained in place until the hemi-

Fontan procedure was performed. The ability to narrow a shunt postoperatively can prove lifesaving in some patients, but the potential risks of thrombosis from its narrowing, and the potential problems with infection from repeated mediastinal exploration for adjustment of the snare, must also be considered.[9]

Pozzi's group[11] described a method of downsizing the shunt using ligaclips (Ethicon Inc., Somerville, NJ). In two patients, this was done intraoperatively, downsizing the shunt from 3.5 to 3.0 mm with good stabilisation of haemodynamics and saturations of oxygen. On four occasions, the shunt was downsized in the intensive care unit 6 to 72 hours postoperatively, with good results. There were no episodes of thrombosis.

Tam et al.[12] used saphenous venous homografts as a systemic-to-pulmonary arterial shunt in 25 patients who underwent the Norwood procedure, hoping to avoid the thrombosis that can occasionally occur with the use of 3-mm shunts. The mean weight of their patients was 3.1 kg, and the mean size of the conduit was 3.2 mm, with 12 patients having shunts that were 3 mm in diameter or smaller. Mortality at 30 days was 8%, and no thrombosis was seen despite the shunts being banded in three patients. They concluded that excellent results could be achieved when the saphenous vein was used to create the shunt. Such channels are less likely to occlude, both acutely and chronically, and permit smaller shunts to be used appropriately in small neonates.

To address the problem of coronary steal during diastole, Sano et al.[13] constructed shunts directly from the right ventricle to the pulmonary arteries, instead of using the classic systemic-to-pulmonary arterial shunt (Fig. 8.6).

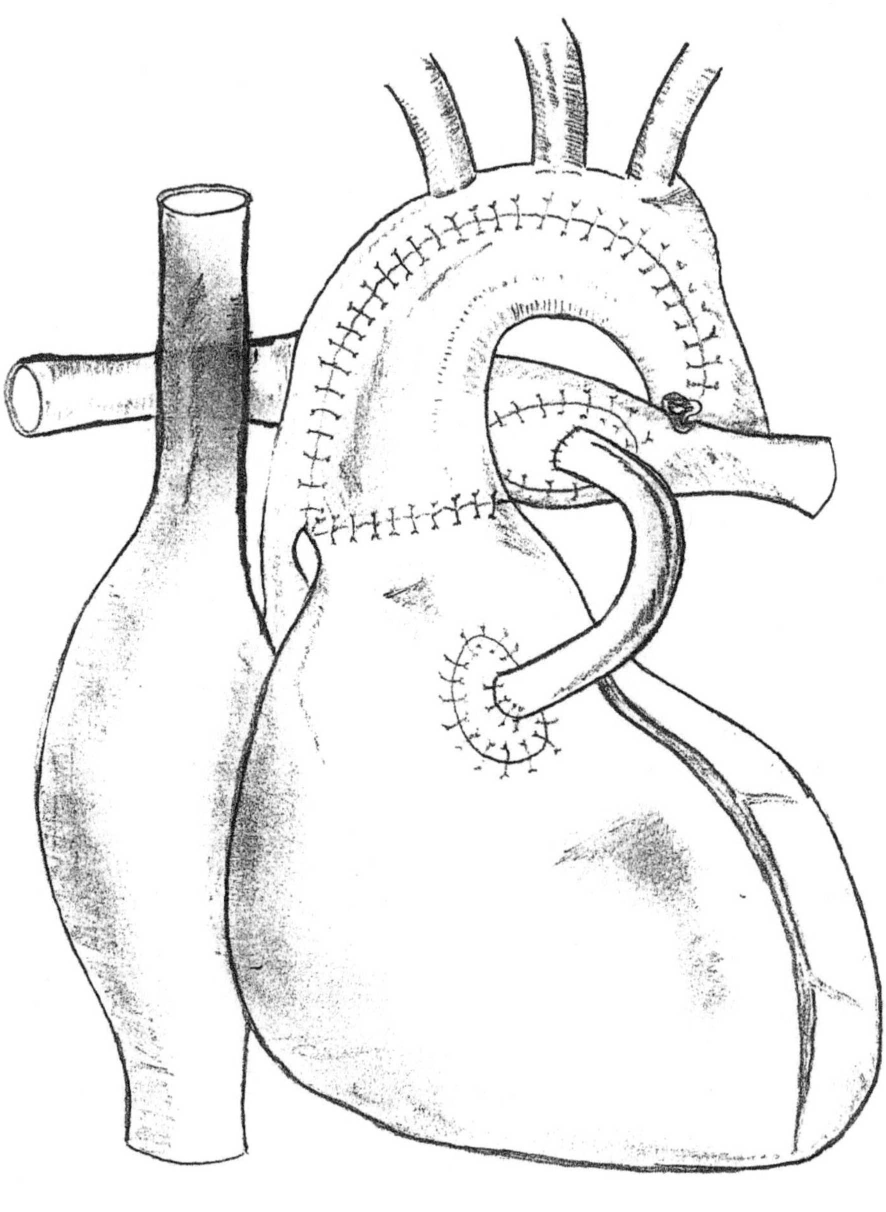

FIGURE 8.6. Diagrammatic representation of the shunt from the right ventricle directly to the pulmonary arteries, as described by Sano and his colleagues.[13]

They reported their experience with 14 patients weighing 1.6 to 3.7 kg and compared them retrospectively with 12 patients in whom a classic Blalock-Taussig shunt was constructed as a part of their Norwood reconstruction. The sizes used were 4 mm in five patients, 5 mm in eight patients, and 6 mm in one patient. Of their patients, 11 survived, including two patients weighing less than 2 kg. Pulmonary overcirculation did not occur in any of the 14 patients. The patients had significantly higher mean diastolic blood pressures than did those in whom classic Blalock-Taussig shunts had been constructed, but at the same time, their tensions of carbon dioxide did not differ. Though the results of Sano et al.[13] look promising, strong reservations about this technique have been expressed by some groups, particularly regarding the effects of ventriculotomy on the function of the morphologically right ventricle, and the long-term issues of arrhythmias from the ventriculotomy scar. We have recently introduced this technique at Alder Hey Children's Hospital. Although our experience is still limited, we have confirmed all the points raised by Sano et al. In particular, the haemodynamic stability in the postoperative period was impressive, being comparable to the typical picture seen in patients following biventricular repair. We also achieved systemic saturations of oxygen in the 90s. We would agree with Sano et al. that with this technique there is preservation of the diastolic pressure, and consequently better preservation of the coronary arterial perfusion. In addition, the coronary arterial and the systemic perfusions are not influenced by changes in the pulmonary vascular resistance. Since a ventricle supports the pulmonary circulation, it can cope much better with changes in the pulmonary vascular resistance. From a surgical point of view, because of the way the conduit is inserted on the pulmonary bifurcation, it maintains a more anatomical position. We speculate that this is likely to reduce the possibility of distortion of the pulmonary arteries. It has been suggested by Sano et al. that because the flow in the conduit occurs only during systole, the conduits do not tend to last as long as the shunts used in the classical Norwood procedure. This has not been our experience.

One obvious question regarding this technique relates to the effects of ventriculotomy on ventricular function. Both in the experience of Sano et al.[13] and in our own experience, the patients who have reached the second stage after this technique have shown echocardiographically and angiographically ventricular function that is akin to that seen in patients coming for the second stage after a conventional Norwood protocol. Norwood's group[14] recently presented their initial results in 12 consecutive patients who underwent the Norwood palliation with a right ventricle to pulmonary artery conduit. The mean weight was 2.9 ± 0.3 kg. All patients had a 5-mm polytetrafluoroethylene tube from the right ventricle to the distal main pulmonary artery. Mean time to extubation among survivors was 34 ± 6 hours, with less than 24 hours in 6 patients. Mean intensive care stay was 6 ± 4 days and mean postoperative stay was 11 ± 4 days. At a mean follow-up of 5.9 ± 2.8 months there was one death out of 12 patients. Preoperative haemodyanamics in 7 patients who underwent the hemi-Fontan procedure showed a mean diastolic pressure of 50 mm Hg, right ventricular end diastolic pressure 9 mm Hg, pulmonary arterial pressure 12 mm Hg, and pulmonary vascular resistance of 1 Wood unit. In spite of being a small group of patients, it can be inferred from this data that a conduit placed between the right ventricle and the pulmonary arteries influences favourably the postoperative physiology of patients undergoing the Norwood palliation. It does not seem to have an adverse effect on the right ventricular function or on the suitability for a future Fontan operation. Clearly a more extensive experience is necessary before a definitive conclusion can be drawn, but there is no doubt in our minds that this technique shows particular promise.

EVOLVING TRENDS IN THE MANAGEMENT OF CARDIOPULMONARY BYPASS

Deep hypothermic circulatory arrest has traditionally been used during the reconstruction of the aortic arch in the Norwood procedure.[15] The safe duration of such arrest has always been a matter of considerable debate. There is evidence to show that brain injury may occur in between one-tenth and one-half of children who have undergone this technique, ranging from subtle but clinically detectable injury to overt and frankly manifest neurodeficits.[16–20] More than one study has demonstrated decreased cognitive ability and motor skills after use of the hypothermic strategy.[20–22] Avoiding deep hypothermic circulatory arrest, therefore, without technically compromising the procedure, intuitively seems the best way of avoiding its potential complications. The last few years have seen the development of techniques for perfusion that cut down or totally avoid the use of circulatory arrest during reparative surgery on the aortic arch. Termed regional low-flow perfusion, this technique may make the use of the deep hypothermic alternative a thing of the past.

Asou et al.[23] were the first group to describe selective cerebral perfusion during repair of the aortic arch in neonates. Cardiopulmonary bypass is initiated by placing a cannula in the pulmonary trunk, and another in the right atrium for venous drainage. During cooling, a polytetrafluorethylene graft is anastomosed to the brachiocephalic artery. The free end of the graft is connected to the arterial perfusion circuit of the cardiopulmonary bypass machine with a Y-shaped connector. At a rectal temperature of 22°C, the arterial cannula is clamped and removed. The brachiocephalic artery just proximal to the graft is clamped, and perfusion to the brain maintained through the graft at a rate of 50 mL/kg/min. After the aortic arch is reconstructed distally, but proximal to the origin of the brachiocephalic artery, the newly reconstructed arch is clamped, establishing perfusion to a now unclamped brachiocephalic artery and descending aorta at a rate of 150 mL/kg/min. The repair is now carried on to the proximal pulmonary trunk and the native ascending aorta to create a systemic outflow tract. After atrial septectomy, perfusion is converted to the neoaorta with the cannula used initially to establish cardiopulmonary bypass. The graft is decannulated and anastomosed to the pulmonary arteries, thus creating a systemic-to-pulmonary arterial shunt.

The technique can be modified to minimise myocardial ischaemic time. The clamp, which usually is placed initially just proximal to the brachiocephalic artery, is instead placed on the arch just distal to the origin of the artery. This allows for myocardial perfusion, as well as perfusion of the brain, while the arch reconstruction is carried out distally. Once reconstruction reaches the level of the clamp, it is moved to a site just proximal to the origin of the graft from the brachiocephalic artery, thus maintaining perfusion to the brain. Cardioplegia is given, the heart is arrested, and the proximal portion of the arch is reconstructed.

Imoto et al.[24] have used combined cannulation of the modified Blalock-Taussig shunt and the descending aorta to perfuse both the upper and lower body during arch reconstruction. Tchervenkov et al.[25] described retrograde perfusion into the brachiocephalic artery via the confluence of the pulmonary arteries and a completed modified Blalock-Taussig shunt, thereby avoiding direct cannulation of either the arch arteries or the shunt. In a more recent update, they described direct cannulation of the distal ascending aorta, which is advanced and snared into place within the brachiocephalic artery once the patient is fully cooled.

Pigula et al.[26] published the first study that quantified cerebral blood flow and cerebral oxygen saturations using regional low-flow perfusion techniques. Using near-infrared spectroscopy, they studied the effects of regional perfusion at low rates of flow on cerebral oxygen saturation and the cerebral blood volume index, in neonates undergoing reconstruction of the aortic arch, comparing the results with similar children who had undergone cardiac repair with deep hypothermic circulatory arrest. Apart from standard monitoring of haemodynamics and temperature, a sensor placed on the right lateral forehead displayed saturations and the blood volume index in real time, which helped direct the rate of flow. With a rate of 20 mL/kg/min, the index and subsequently the saturations of cerebral perfusion were restored to baseline levels as obtained on full cardiopulmonary bypass. In the control group, there was a prompt decline in both the cerebral index and saturations upon initiation of deep hypothermic circulatory arrest. Data suggested that, for a neonate, flows of 20 mL/kg/min using the alpha-stat strategy for management of blood gases provides adequate circulatory support.

In a follow-up study, the same group set out to determine the ability of such regional perfusion at low rates of flow to provide subdiaphragmatic circulatory support, having been impressed with the copious back bleeding from the descending aorta upon removal of the clamp during their earlier experience with the technique.[27] They again studied neonates undergoing reconstruction of the aortic arch. Three techniques were used to assess somatic perfusion: abdominal aortic blood pressure, flow of blood through the quadriceps muscle by near-infrared spectroscopy, and gastric tonometry. They found that abdominal aortic blood pressure was higher and the quadriceps blood volumes and oxygen saturations were greater during regional perfusion at low rates of flow than during deep hypothermic circulatory arrest. During rewarming, the carbon dioxide tension difference between arterial blood and gastric mucosa was lower after deep hypothermic circulatory arrest than after regional low flow perfusion. They concluded that regional perfusion provides significant subdiaphragmatic circulatory support during reconstruction of the arch. Morbidity and mortality in patients who are haemodynamically stable after a Norwood procedure generally results from sepsis or multiorgan failure, which are events that often have their roots in subdiaphragmatic organ dysfunction, mainly of the kidneys, gut, or liver. It would seem prudent to believe that maintenance of subdiaphragmatic perfusion during reconstruction of the arch would help maintain a milieu that would be favourable to the recovery of these precariously sick patients.

APPLICATION OF THE NORWOOD PRINCIPLE TO CARDIAC MALFORMATIONS OTHER THAN HYPOPLASIA OF THE LEFT HEART

The principles of Norwood operation can be applied to a variety of complex cardiac malformations other than the classic cases with hypoplasia of the left heart.[28] This includes lesions with the physiological features of obstruction to the aortic outlet and ductal-dependent systemic circulations. Initial palliation involves creation of an unobstructed aortic outflow tract, incorporating a proximal anastomosis between the pulmonary arteries and the aorta, with or without patch augmentation of the aorta. Flow of blood to the lungs is reestablished by creating a systemic-to-pulmonary arterial shunt, in a manner analogous to the first-stage Norwood procedure. The subsequent treatment in these patients depends on the adequacy of size and function of the two ventricles. While the group with a functionally single ventricle predictably goes down the Fontan pathway, a select group with two functioning ventricles can be converted to biventricular repair within the first year of life. In centres with expertise in neonatal surgery, a one-stage biventricular repair, combining Norwood-type reconstruction of the aortic arch with closure of the ventricular septal defect to the pulmonary trunk and placement of a conduit from the right ventricle to the pulmonary arteries, is done for cardiac malformations like interrupted aortic arch with ventricular septal defect and obstruction to the left ventricular outflow tract. The anatomical substrates in those with functionally single ventricles other than hypoplasia of the left heart that are amenable to the Norwood type of repair include double inlet left ventricle or tricuspid atresia with discordant ventriculoarterial connections. Anatomical substrates that can be converted to an eventual biventricular repair incorporating the Norwood principle include hypoplasia or interruption of aortic arch with ventricular septal defect and obstruction to the left ventricular outflow tract, and double outlet right ventricle with subaortic obstruction.

CONCLUSION

The techniques to manage hypoplasia of the left heart have continued to evolve, even in the face of a high mortality in the initial days of the Norwood procedure. From a mortality rate in excess of 50% at the beginning, the procedure is now performed at some centres with a mortality of 15% or less, a remarkable achievement indeed. Additionally, an improved understanding of the physiology created by the Norwood procedure and the increasing experience with reconstruction of the aortic arch have significantly improved our management of patients with functionally univentricular hearts other than hypoplasia of the left heart. The Norwood procedure is emerging as a Norwood principle, which is being increasingly employed to achieve biventricular repairs in neonates and infants with a combination of interruption or hypoplasia of the aortic arch, obstruction of the left ventricular outflow tract, ventricular septal defect, and two good ventricles.

References

1. Norwood WI, Lang P, Hansen DD. Physiologic repair of aortic atresia-hypoplastic left heart syndrome. N Engl J Med 1983;308:23–26.
2. Bu'Lock FA, Stumper O, Jagtap R, et al. Surgery for infants with a hypoplastic systemic ventricle and severe outflow obstruction: early results with a modified Norwood procedure. Br Heart J 1995;73:456–461.
3. Fraser CD, Mee RBB. Modified Norwood procedure for hypoplastic left heart syndrome. Ann Thorac Surg 1995;60(suppl):S546–549.
4. Mahle WT, Rychick J, Weinberg PM, Cohen MS. Growth characteristics of the aortic arch after the Norwood operation. J Am Coll Cardiol 1998;32:1951–1954.
5. Fraisse A, Colan SD, Jonas RA, Gauvreau K, Geva T. Accuracy of echocardiography for detection of aortic arch obstruction after stage I Norwood procedure. Am Heart J 1998;135(2 pt 1):230–236.
6. Ishino K, Stumper O, De Giovanni JJV, et al. The modified Norwood procedure for hypoplastic left heart syndrome: early to intermediate results of 120 patients with particular reference to aortic arch repair. J Thorac Cardiovasc Surg 1999;117:920–930.
7. Bartrum U, Grunenfelder J, Van Praagh R. Causes of death after the modified Norwood procedure: a study of 122 post-mortem cases. Ann Thorac Surg 1997;64: 1795–1802.
8. Poirier NC, Drummond-Webb JJ, Hisamochi K, Imamura M, Harrison AM, Mee RBB. Modified Norwood procedure with a high flow cardiopulmonary bypass strategy results in low mortality without late arch obstruction. J Thorac Cardiovasc Surg 2000;120:875–884.
9. Pearl JM, Nelson DP, Schwartz SM, Manning PB. First-stage palliation for hypoplastic left heart syndrome in the twenty-first century. Ann Thorac Surg 2002;73:331–340.
10. Schmid FX, Kampmann C, Kuroczynski W, et al. Adjustable torniquet to manipulate pulmonary blood flow after Norwood operations. Ann Thorac Surg 1999;68:2306–2309.
11. Kuduvalli M, Mclaughlin KE, Trivedi DB, Pozzi M. Norwood type operation with adjustable systemic to pulmonary artery shunt using haemostatic clip. Ann Thorac Surg 2001;72(2):634–635.
12. Tam VK, Murphy K, Parks WJ, et al. Saphenous vein homograft: a superior conduit for the systemic arterial shunt in the Norwood operation. Ann Thorac Surg 2001;71:1537–1540.
13. Sano S, Ishino K, Kawada M, et al. Right ventricle-pulmonary artery shunt in first stage palliation for hypoplastic left heart syndrome. J Thorac Cardiovasc Surg 2003;126:504–509.
14. Pizarro C, Malec M, Mather KO, et al. Right ventricle to pulmonary artery conduit improves outcome after stage I Norwood for hypoplastic left heart syndrome. Circulation 2003;108(Suppl I):II:155–160.

15. Brunberg JA, Reilly EL, Doty DB. Central nervous system consequences in infants of cardiac surgery using deep hypothermia and circulatory arrest. Circulation 1974;50(suppl):II60–II68.
16. Clarkson PM, MacArthur BA, Barrat-Boyes BG, et al. Developmental progress following cardiac surgery in infancy using profound hypothermia and circulatory arrest. Circulation 1980;62:855–861.
17. Ehyai A, Fenichel GM, Bender HW. Incidence and prognosis of seizures in infants after cardiac surgery with profound hypothermia and circulatory arrest. JAMA 1984;252:3165–3167.
18. Jonas RA, Wernovsky G, Ware GD, et al. The Boston Circulatory Arrest Study: perioperative neurologic and developmental outcome after arterial switch operation. Circulation 1992;86(suppl):I360.
19. Hickey PR. Neurologic sequels associated with hypoplastic left heart syndrome. Ann Thorac Surg 1998; 65:S65–70.
20. Jonas RA. Hypothermic circulatory arrest and the paediatric brain. J Cardiothorac Vasc Anesth 1996;10:66–74.
21. Kern JH, Hinto VJ, Nereo NE, et al. Early developmental outcome after the Norwood procedure for hypoplastic left heart syndrome. Paediatrics 1998; 102:1148–1152.
22. Uzark K, Lincoln A, Lamberti JJ, et al. Neurodevelopmental outcomes in children with Fontan repair of functional single ventricle. Paediatrics 1998;101(4 pt 1): 630–633.
23. Asou T, Kado H, Imoto Y, et al. Selective cerebral perfusion technique during aortic arch repair in neonates. Ann Thorac Surg 1996;61:1546–1548.
24. Imoto Y, Kado H, Shiokawa Y, et al. Norwood procedure without circulatory arrest. Ann Thorac Surg 1999;68: 559–561.
25. Tchervenkov CI, Korkola SJ, Shum-Tim D. Surgical technique to avoid circulatory arrest and direct arch vessel cannulation during neonatal aortic arch reconstruction. Eur J Cardiothorac Surg 2001;19:708–710.
26. Pigula FA, Nemoto EM, Griffith BP, Siewers RD. Regional low flow perfusion provides cerebral circulatory support during neonatal arch reconstruction. J Thorac Cardiovasc Surg 2000;119:331–339.
27. Pigula FA, Gandhi SK, Siewers RD, et al. Regional low flow perfusion provides somatic circulatory support during neonatal arch surgery. Ann Thorac Surg 2001;72:401–407.
28. Jacobs ML, Rychik J, Murphy JD, et al. Results of Norwood operation for lesions other than hypoplastic left heart syndrome. J Thorac Cardiovasc Surg 1995; 110:1555–1562.

Chapter 9

ANAESTHESIA FOR HYPOPLASIA OF THE LEFT HEART

Monica A. Stokes

The anaesthetist has a central role in the perioperative management of infants with hypoplasia of the left heart, contributing to preoperative stabilisation and transport of the newborn, as well as to the events surrounding surgery and postoperative care. He or she contributes to the team approach to treatment, and is well placed to observe and understand the roles of other team members and the difficulties they may encounter.

This chapter reviews the essential features of pathophysiology that have a bearing on perioperative management, highlights aspects of monitoring and equipment that are considered to be important at Birmingham Children's Hospital, and describes the general anaesthetic management used at this hospital. The views expressed are based on my personal experience and reflect our local practice, but I hope they will be of value to those supporting similar surgical programmes.

THE PARALLEL CIRCULATIONS

The pathophysiological effects of the defects associated with hypoplasia of the left heart, typically aortic and mitral valvular stenosis or atresia along with hypoplasia of the ascending aorta and left ventricle, are similar despite the considerable variation in anatomic combinations.[1] The left ventricle is nonfunctional. Pulmonary and systemic venous blood, therefore, must mix at the atrial level through a septal defect, a stretched oval foramen, or rarely through malformations such as the levoatrial cardinal vein or anomalous pulmonary venous connections when the atrial septum is intact. Systemic and pulmonary circulations are then supplied in a parallel fashion from the morphologically right ventricle, with systemic flow dependent on patency of the arterial duct. More explicitly, perfusion to the transverse aortic arch and ascending aorta is retrograde through the duct, and perfusion to the lower body is anterograde through the descending aorta. The important feature of this circulatory pattern is that, because the pulmonary arteries are connected in parallel with the duct and descending aorta, the proportion of flow through each depends on the balance between pulmonary and systemic vascular resistances. Untreated, this malformation is lethal. The natural tendency for the duct to close results in impaired systemic and coronary arterial perfusion, metabolic acidosis, and ischaemia. Even if the duct remains patent in some neonates, the normal physiological decrease in pulmonary vascular resistance produces an increase in the flow of blood to the lungs that, in this arrangement with parallel circulations, is at the expense of systemic flow. The result, again, is acidosis, inadequate systemic and coronary perfusion, and death despite an improved arterial oxygenation.

PREOPERATIVE PREPARATION

Most infants present for surgery in the first week of life. The diagnosis may have been made on an antenatal ultrasonic scan, allowing time for parental counselling and for arrangements to be made for the baby to be delivered close to the centre for paediatric cardiac surgery. If the diagnosis is made after birth, the baby is transferred following resuscitation and parental counselling. Accurate and prompt diagnosis is essential, because well-intentioned but inappropriate early management, such as hyperventilation with 100% oxygen, can worsen rather than improve metabolic acidosis.

Good communication with one's cardiological colleagues provides insight into the essential features of the echocardiographic diagnosis that will become relevant postoperatively. One of these is the degree, if any, of tricuspid valvular regurgitation. When severe, this is unlikely to be compatible with a successful surgical reconstruction. Another is the degree of aortic hypoplasia, and the adequacy of the interatrial communication. Although routinely performed, chest radiographic findings of cardiomegaly and increased pulmonary vascular markings are supportive of the diagnosis rather than confirmatory, as is the evidence of right atrial and ventricular hypertrophy on electrocardiogram.

The most important preoperative intervention is to commence infusions of prostaglandin E_1, known as Prostin in the United Kingdom, through a reliable vein, to maintain patency of the arterial duct. Further management depends on the relative balance between pulmonary and systemic blood flow in individual patients (Table 9.1). Those with a relatively restrictive interatrial communication are likely to have adequate flow of blood to the lungs, and mild pulmonary venous hypertension without pulmonary congestion. With the duct patent, these babies are often warm and well perfused, with no evidence of hypotension or metabolic acidosis. In this situation, surgery can be scheduled electively. Unfortunately, it does not take much to upset this balance. The timing of surgery, while not urgent, takes account of this.

Some babies require admission to the intensive care unit for ventilation because Prostin has caused apnoeic pauses, but it is debatable whether all should routinely be ventilated and "lined-up" in the intensive care unit to await surgery. The effects of these interventions may be other than intended. Most intravenous hypnotic drugs cause a fall in systemic vascular resistance, and manipulation of the arterial tension of carbon dioxide with ventilation alters the pulmonary vascular resistance. When preoperative ventilation is necessary, our practice is to give ketamine intravenously as a hypnotic, at a dose of 1.2 mg/kg, followed by infusions of morphine and rocuronium for sedation and muscular paralysis, respectively. It is important that the airways be managed securely, so endotracheal intubation is either carried out or closely supervised by a senior clinician. It is very easy for overenthusiastic ventilation with a high concentration of inspired oxygen to disrupt the balance of this circulation, favouring flow of blood to the lungs at the expense of the systemic circulation. Similarly, the endotracheal tube should be securely fixed, because accidental extubation and its sequels are poorly tolerated.

Where the interatrial communication is large and unrestrictive, there is unimpeded flow of blood to the lungs, and systemic hypoperfusion. Attempts by the baby to compensate for the resultant metabolic acidosis by hyperventilating only worsen the acidosis, because the effect of hyperventilation is to lower pulmonary vascular resistance, further prejudicing systemic flow. These babies need endotracheal intubation and ventilation, and careful management of arterial blood gases to maintain an elevated pulmonary vascular resistance. We aim for arterial tensions of carbon dioxide at the high side of the normal range, and deliver a low concentration of inspired oxygen, often 21%, accepting values of saturations of haemoglobin between 80% and 85%. The danger with manipulating tidal volume and ventilatory rate in pursuit of "ideal" blood gases is the risk of serious hypoventilation, atelectasis, and hypoxaemia secondary to collapse of the lungs. This is probably a greater danger than tolerating higher than ideal saturations of haemoglobin in a baby awaiting surgery who is otherwise well perfused and not acidotic. To date, we have not found the need to add supplemental carbon dioxide to inspired gases.[2,3] Good sedation is essential because necessary, but potent, stimuli such as endotracheal suctioning can provoke desaturation and haemodynamic instability. Such stimuli can be anticipated, and sedation deepened, as required.[4]

Another pitfall is inadvertent hypovolaemia, and most infants benefit from intermittent boluses of 4.5% albumin at a dose of 10 mL/kg as indicated. The guidelines are provided by clinical assessment of systemic perfusion, with warm limbs and a wet diaper being good signs. Approximately one-third of infants with hypoplasia of the left heart undergoing surgical palliation in our unit require preoperative inotropic support, and we give dobutamine at a dose of 5 to 10 μg/kg/min, usually because they have been rescued from an episode of ductal closure. Resuscitation should be completed and the baby stabilised prior to attempting surgical pallia-

TABLE 9.1. Factors affecting the haemodynamic equilibrium in patients with hypoplasia of the left heart.

Manipulations to increase pulmonary ventricular resistance	*Manipulations to decrease pulmonary ventricular resistance*
Decreased inspired concentration of oxygen	Increased inspired concentration of oxygen
Hypoventilation, with arterial tensions of carbon dioxide from 40 to 50 mm Hg	Hypoventilation, with arterial tensions of carbon dioxide from 20 to 25 mm Hg
Positive end-expiratory pressure	Inotropic support

tion. Rarely, the anatomy may produce insufficient mixing at atrial level, poor flow of blood to the lungs, and profound hypoxaemia. This is the exception to the rule that surgery is a scheduled activity.

The success of the first stage of palliation, in which there is ligation of the patent duct, atrial septectomy, division of the pulmonary trunk, aortic reconstruction, and establishment of a systemic-to-pulmonary arterial shunt, does not alter the underlying parallel arrangement of the circulatory patterns. Rather than being duct-dependent, the pulmonary circulation is now shunt dependent, and systemic perfusion is still dependent on the relative balance of pulmonary and vascular resistances.

EQUIPMENT AND MONITORING

Transfers of the patient should be treated as an extension of intensive care, with the same attention to detail to maintain physiological balance. The pitfalls we have encountered relate to provision of a constant arterial concentration of oxygen and ventilatory pattern; avoidance of hyperventilation during manual ventilation, with subsequent swings in pH and arterial tension of carbon dioxide; and problems with battery-powered infusion devices.[5] These are easily overcome, once identified, and full facilities for monitoring during transport, even for short distances, are invaluable. Cylinders of both air and oxygen are required.

With experience, we have revised our views on the placement of central lines, not only for children with hypoplasia of the left heart but also for all those with functionally univentricular physiology whose treatment is directed toward the Fontan procedure. For most open-heart surgery, our preference is to place a 4- or 5-French triple-lumen catheter percutaneously via the internal jugular vein. Some patients, however, may have developed calcification or thrombus in the superior caval vein, the causes of which are probably multifactorial, but occur in the context of infusion of highly concentrated drugs under conditions of low cardiac output. Such a complication is immediately life-threatening if the thrombus is dislodged and occludes the Blalock-Taussig shunt, or can seriously prejudice a successful second-stage cavopulmonary shunt if it results in elevated pulmonary vascular pressure. We now choose to place triple-lumen catheters via the femoral vein, or the umbilical vein using a Seldinger technique, and we confirm the position of the catheter using ultrasound. The disadvantage of routine use of the femoral vein in this context is that scarring and thrombosis may make subsequent vascular access for cardiac catheterisation more difficult. On balance, nonetheless, the risks of serious morbidity appear lower. If it is not possible to place a central venous line percutaneously, it is important not to struggle, risking hypothermia, hypovolaemia, and acidosis. The alternative is for the surgeon to place a catheter directly into the atrium during the operation, bringing it out through the surgical field.

We remove electively any umbilical arterial catheter that may have been placed by a referring hospital, once an alternative site of arterial pressure monitoring has been secured. Although these are easy to place in the newborn, there are concerns of an increased risk of ischaemia of the legs and an association with necrotising enterocolitis.[6] Fortunately, when the duct is patent, neonates with hypoplasia of the left heart usually have easily palpable peripheral pulses. One then has the luxury of debating the optimal site for arterial cannulation. Some authorities suggest placing a left radial arterial line, because pressures monitored from the right arm may be misleadingly low after formation of a right-sided Blalock-Taussig shunt. Given that coarctation may accompany the hypoplastic aorta,[7] my own preference is to place lines in both the right radial and femoral arteries for the duration of surgery. The radial line can be removed early in the postoperative period, but the additional information obtained from comparative study of the waveforms in the operating room can be very helpful if surgical reconstruction of the aorta proves difficult. Similarly, placing an additional pulse oximeter on the patient's foot acts as a monitor of perfusion, giving a reassuring sign, regardless of the value of the reading, that there is good distal blood flow. Pulse oximeters that automatically amplify a poor signal should be treated with caution. Otherwise, intraoperative monitoring is similar to any other neonatal procedure with cardiopulmonary bypass, requiring the recording of the electrocardiogram; arterial levels of oxygen; nasal, oesophageal, and skin temperatures; and capnography. Although the recorded values of end-tidal carbon dioxide will not accurately reflect the arterial tension of carbon dioxide in the presence of an intracardiac shunt, and there is no substitute for regular measurements of arterial blood gases, the capnograph is useful as a detector of flow of blood to the lungs once the systemic-to-pulmonary arterial shunt has been constructed, and preparations are made to separate from bypass.

INTRAOPERATIVE MANAGEMENT

We find ketamine, at a dose of 1 to 2 mg/kg, to be a well-tolerated agent for the induction of anaesthesia, both for the first stage of palliation and for other procedures in infants with hypoplasia of the left heart. Subsequently, and when the infant is transferred from the intensive care unit already intubated and sedated, a surgical plane of anaesthesia is provided using boluses or infusion of fentanyl, titrated to haemodynamic responses and supplemented by inhalation of low concentrations of isoflurane. Muscular paralysis is controlled by intermittent boluses of pancuronium, or infusion of rocuronium, and ventilation is adjusted by controlling pressures to achieve the same arterial blood gases as on the intensive care unit. All babies are intubated nasotracheally, producing a secure and comfortable route for postoperative care, using an endotracheal tube cut to the appropriate length. Keeping the ambient temperature warm at this stage is helpful if additional vascular access has to be secured. Once the patient and surgeon are both positioned and prepared, the temperature can be allowed to drift downward, as surgery involves a period of deep hypothermic circulatory arrest. Routine antibiotic prophylaxis is given, and Prostin is infused until the start of cardiopulmonary bypass. Individual neonates vary in their requirements for fentanyl. In general, nonetheless, this anaesthetic approach produces a stable pre-bypass phase. Small increments of colloid may be needed.

During cardiopulmonary bypass, administration of isoflurane via the pump has proved helpful in controlling systemic vascular resistance, particularly on rewarming. At this stage, vasodilation is usually supplemented by sodium nitroprusside at a dose of 1 to 2 μg/kg/min. Although some neonates are quite sensitive to the vasodilatory effects of sodium nitroprusside, its short half-life gives sensitive control during rewarming and separation from bypass. We have recently added phenoxybenzamine to this strategy because its action in producing alpha-blockade, coupled with its prolonged receptor-binding characteristics, appear to be beneficial in promoting stability of pulmonary vascular resistances.[8] A single dose of 1 mg/kg is given on commencement of cardiopulmonary bypass via the cardiopulmonary bypass circuit. We have found that all patients benefit from inotropic support, and our usual practice has been to commence dobutamine at a level of 10 μg/kg/min, along with an infusion of adrenaline at a concentration of between 0.01 and 0.05 μg/kg/min as the heart is reperfused, adjusting these as indicated once separated from bypass. With judicious titration, excessive peripheral vasoconstriction can be avoided. More recently, we have started substituting milrinone for dobutamine in this strategy, supplemented when necessary by adrenaline in low dose. Having corrected any acidosis or electrolytic disturbance, and with a haematocrit of about 35% to 38%, we separate from bypass, ventilating with an inspired concentration of oxygen of 50%, and we observe. Characteristically, the first 20 minutes or so off bypass is a time of pressure and saturation flux, as pulmonary and systemic vascular resistances equilibrate. Our approach is to ventilate to achieve an arterial tension of carbon dioxide of approximately 40 to 45 mm Hg in the first instance and reduce the inspired concentration of oxygen gradually, accepting saturations of haemoglobin in the range of 75% to 80%, or thereabouts. Before lower saturations are blamed on an insufficient or partially obstructed shunt, the more obvious causes of an increased pulmonary ventricular resistance immediately subsequent to bypass should be investigated. These are inspissated bronchial secretions and lung collapse, and inadequate rewarming. While the surgeon is correcting these, the patient can be supported by careful replacement of volume, adjustment of inotropes, and increased ventilation. A saturation of haemoglobin greater than 85% indicates that too much blood is flowing to the lungs, and requires urgent attention. The level of ventilation should be adjusted to increase the pulmonary ventricular resistance, while the surgical team confirms that there is no obstruction to aortic flow.

With time, inotropic support, and calcium administration, any systemic hypoperfusion due to myocardial dysfunction subsequent to bypass should improve. Persistent myocardial ischaemia, however, is an indication to review the surgery! Volume is replaced using common atrial and systemic pressures as a guide. Although it is impossible to be dogmatic, systolic pressures of 70 to 75 mm Hg, and atrial pressures between 7 and 9 cm H_2O, are usually satisfactory at this stage. Temporary occlusion of the shunt and observation of the change produced in diastolic pressure may give a feel for pulmonary ventricular resistance. Blood, platelets, and fresh frozen plasma are transfused according to haemodynamic responses, aiming for an haematrocrit of about 38%, and ionised calcium, serum potassium, and acid-base balance are carefully controlled. We have a low tolerance of metabolic acidosis, and will correct a base deficit of 5 mM or greater, and actively pursue the reason for a persistent or increasing metabolic acidosis. When measured, we typically find that, at this stage, the levels of lactate in the serum are higher when compared with other neonates undergoing cardiopulmonary bypass. Values of 8 to 12 mM are not uncommon, but we would expect to see the concentration of lactate in the serum to fall over the subsequent 24 hours. Full monitoring is continued during transfer to the intensive care unit, and none of the supportive drug infusions is interrupted.

A detailed discussion of postoperative intensive care management is outside the scope of this chapter. In general, we aim for delayed sternal closure, and all infants receive infusions of morphine for deep sedation, using rocuronium for muscular paralysis until then.

Although more easily said than achieved, the aim is to limit physiological instability due to marked changes in the production of carbon dioxide and the demand for oxygen. Infusions of nitroprusside and adrenaline are weaned, usually during the first 4 to 8 hours, during which the pulmonary and systemic vascular resistances may demonstrate considerable fluctuations, but dobutamine is continued until ventilatory weaning and extubation are achieved. A persistently elevated level of lactate in the serum, particularly if higher than 5 mmol/L, is a general indication of poor cardiac output. Other indirect monitors of this complication are the quality of the peripheral pulses, capillary refill time, urinar output, and base deficit. Some have found continuous assessment of saturations of oxygen in the superior caval vein helpful as an approximation of mixed venous saturation, giving warning of reductions in cardiac output before arterial desturation and blood pressure are affected.[9,10]

ANAESTHESIA FOLLOWING THE FIRST STAGE OF PALLIATION IN THE NORWOOD SEQUENCE

Children who have successful completed the first stage of palliation require anaesthesia for cardiac catheterisation before the second stage, the cavopulmonary anastomosis, and ultimately completion of the Fontan circulation. At any age, a child with hypoplasia of the left heart may also need to be anaesthetised for noncardiac surgical procedures. It is important, therefore, that careful consideration be given to the environment and experience of staff involved in providing this service.

Although it is our experience that the survivors of the first stage of palliation tolerate subsequent procedures with a greater degree of stability than in the neonatal period, these children remain in tenuous physiological balance. In particular, changes in intravascular volume are poorly tolerated, and care should be taken to avoid excessive periods of preoperative starvation, and to anticipate and treat hypovolaemia promptly. The clinical history may give some indication about likely anaesthetic difficulties: increasing hypoxaemia, for instance, that may result because of decreased perfusion through the shunt, or because a concurrent respiratory infection may be exacerbating function. The morphologically right ventricle supporting the systemic circulation can show signs of failure early if there is any obstruction to aortic flow, so consultation with cardiological colleagues is very helpful before embarking on procedures for which detailed preoperative cardiovascular assessment is not usually performed. Antibiotic prophylaxis against endocarditis is essential. Although many children appear to grow and develop well, formal neuropsychiatric and neurodevelopmental testing may reveal abnormalities.[11] The traumas of repeated operations can be reduced by a sympathetic approach to the whole family, and by the judicious use of sedative premedication before induction of anaesthesia.

References

1. Noonan JA, Nadas AS. The hypoplastic left heart syndrome: an analysis of 101 cases. Paediatr Clin North Am 1958;5:1029–1056.
2. Shime N, Hashimoto S, Hiramatsu N, et al. Hypoxic gas therapy using nitrogen in the preoperative management of neonates with hypoplastic left heart syndrome. Paediatr Crit Care Med 2000;1:38–41.
3. Jobes DR, Nicolson SC, Steven JM, et al. Carbon dioxide prevents pulmonary overcirculation in hypoplastic left heart syndrome. Ann Thorac Surg 1992;54:150–151.
4. Hickery PR, Hansen DD. High-dose fentanyl reduces intraoperative ventricular fibrillation in neonates with hypoplastic left heart syndrome. J Clin Anaesth 1991;3:295–300.
5. Tatman A, Brunner H, Stokes M. Premature failure of battery-powered pumps. Anaesthesia 1996;51: 1041–1042.
6. McElhinney DB, Hedrick HL, Bush DM, et al. Necrotising enterocolitis in neonates with congenital heart disease: risk factors and outcomes. Paediatrics 2000;106:1080–1087.
7. Machii M, Becker AE. Nature of coarctation in hypoplastic left heart syndrome. Ann Thorac Surg 1995;59:1491–1494.
8. Tweddell JS, Hoffman GM, Mussatto KA, et al. Improved survival of patients undergoing palliation of hypoplastic left heart: lessons learned from 115 consecutive patients. Circulation 2002;106:1–82.
9. Butheitel G, Scharf J, Hofbeck M, et al. Estimation of cardiac index by means of the arterial and the mixed venous oxygen content and pulmonary oxygen uptake determination in the early post-operative period following surgery of congenital heart disease. Intensive Care Med 1994;20:500–503.
10. Rogers BT, Msall ME, Buck GM, et al. Neurodevelopmental outcome of infants with hypoplastic left heart syndrome. J Paediatr 1995;126:496–498.
11. Mahle WT, Clancy RR, Moss EM, et al. Neurodevelopmental outcome and life-style assessment in school-aged and adolescent children with hypoplastic left heart syndrome. Paediatrics 2000;105:1082–1089.

Chapter 10

TRANSPLANTATION IN THE MANAGEMENT OF INFANTS WITH HYPOPLASIA OF THE LEFT HEART

Anees J. Razzouk, Richard E. Chinnock, Joyce K. Johnston, James A. Fitts, and Leonard L. Bailey

Hypoplasia of the left heart encompasses a relatively constant spectrum of congenital cardiac malformations (see Chapter 1). Abnormalities of other systems or organs are rare. Except for their heart, most infants with this malformation would otherwise be normal. The constellation of lesions is numerically significant, and accounts for almost one-tenth of the congenital cardiac anomalies producing symptoms in the first year of life. Its most appropriate treatment, therefore, is also of major significance.

HISTORICAL ASPECTS

The past decades witnessed a dramatic evolution in the clinical treatment of infants born with this constellation of lesions. With the introduction of staged reconstruction following the Norwood sequence and with the availability of transplantation, the lesion is no longer necessarily fatal.[1–6] Yacoub first attempted allotransplantation in July 1984, although the experience was not formally reported. The infant died $2\frac{1}{2}$ weeks later of multiple complications. Shortly thereafter, in October 1984, based on extensive experimental work on newborn goats, Bailey and his colleagues[7] implanted a selected baboon heart into a 12-day-old girl, now known as Baby Fae. Function of the graft was normal during the first 14 days but the baby died later of disseminated intravascular coagulopathy stimulated by an immunological mismatch between the donor organ and the recipient. A year later, in November 1985, Bailey performed the first successful newborn cardiac allotransplantation. The 4-day-old male recipient received the new heart at Loma Linda University Medical Center, Loma Linda, California, and remains alive and well 18 years later.

Cardiac transplantation for hypoplasia of the left heart offers the promise of virtually normal cardiovascular physiology. Such therapy has now become a clinical reality limited only by the availability of donor hearts.

The material presented in this chapter is based on the experience at Loma Linda University with over 151 primary allotransplants for hypoplasia of the left heart. Of these 151 patients, 95 were male and 56 were female. Their age at the time of transplantation ranged from 1 day to 6 months. The average weight at the time of transplantation was 3.7 kg, with a range from 2.05 to 6.43 kg. The discussion focuses on the technical and practical aspects of this therapy.

MANAGEMENT PRIOR TO TRANSPLANTATION

The shortage of donor organs has contributed to the long waiting time for infants requiring transplantation. The waiting period in our series, from birth to transplantation, ranged from 1 to 183 days, and averaged 40 days, with a variance of 34 days. Of listed infants, 23% died while waiting. Death prior to transplantation is institution-dependent, and can be reduced, and, it is hoped, one day eliminated, by expanding the pool of donors and improving the management of infants waiting for transplantation.

In the past, when the diagnosis of hypoplasia of the left heart was made during fetal life, the fetus was registered and became a potential candidate for transplantation at 36 weeks' gestation. Surgical delivery was performed thereafter should a donor organ become available; otherwise the pregnancy was carried to term. In our series, two of 32 fetuses registered during fetal life were delivered by caesarean section, and transplanted at

1.5 and 3 hours of life.[8] More recently, we have abandoned our protocol for fetal listing because of the potential for other significant unexpected anatomic abnormalities.[9]

More commonly, the diagnosis is made after birth, and not infrequently after the baby has been discharged home. Spontaneous closure of the arterial duct, several days after birth, results in severe cardiovascular collapse. Aggressive resuscitation of those infants is often successful with fluids, correction of acidosis, infusion of prostaglandin E_1, mechanical ventilation, and delivery of inotropes at low dosage. Medical support, except for prostaglandin E_1, is gradually weaned once the infant reaches cardiorespiratory stability. The major challenge during the period prior to transplantation is the maintenance of satisfactory systemic perfusion, and achieving a balance between the pulmonary and systemic vascular resistances. We managed several infants on prostaglandin E_1, who waited more than 6 months from birth until an organ became available. The side effects of prostaglandin E_1, namely apnoea, fever, and cortical hyperostosis, are dose-related. The dose, therefore, should be kept at the minimal level necessary to maintain ductal patency, usually from 0.0125 to 0.05 μg/kg/min. Ductal closure commonly results from interruption of the infusion of prostaglandin, and rarely from resistance to the drug. Periodic echocardiography is essential during the period of waiting to confirm unrestricted flow across the duct, and to establish the size of the interatrial communication. Infants with signs of systemic hypoperfusion and early ductal closure may benefit from percutaneous placement of a ductal stent. In our series, eight infants had a catheter-mounted stent placed in the duct, and four were successfully transplanted.[10]

Keeping the infants on room air, and avoiding hyperoxia, maintaining saturations of oxygen at between 70% and 85%, and similarly avoiding hypocapnia, best achieve balancing the distribution of blood between the pulmonary and systemic circuits. Some infants with intractable pulmonary oedema and saturations of oxygen greater than 90% may benefit from diuretics and reduced administration of oxygen.

Table 10.1. Interventions prior to transplantation: the experience at Loma Linda University, Lorma Linda, California

Indication	*Procedure*	*Number*
Stenosing or closing duct resistant to prostaglandin E_1	Ductal stent	8
Restrictive interatrial communication	Balloon septostomy	45
	Blade septostomy	1
	Open septectomy	8

On occasion, hypoxemia, or severe oxygen desaturation, as a consequence of a low ratio of pulmonary to systemic flows, may be due to closure or severe restriction of the interatrial communication. Such infants are at high risk for death while awaiting transplantation, and require intervention to enlarge the atrial septal defect (Table 10.1).

An adequate nutritional state of the recipients is generally maintained with enteral feedings. Rarely is parenteral support needed. Blood transfusions with washed and irradiated or leukocyte-filtered red blood cells may be indicated should levels of haemoglobin fall to less than 10 g/dL.

Using these techniques of management, neonatologists, along with a team of specialised medical and surgical consultants, have been successful at supporting infants for up to 6 months. Along with the special care and monitoring of the infant, the family receives psychosocial preparation and the proper teaching that is essential for satisfactory long-term outcome.

Cardiac evaluation prior to transplantation includes an electrocardiogram, chest x-ray, and an echocardiogram (see Chapter 5). The classical anatomical features are very well demonstrated on echocardiography, so much so that prenatal diagnosis is now possible before 20 weeks' gestation. This limits the need for cardiac catheterisation or magnetic resonance imaging to rare situations where other complex congenital malformations are present along with hypoplasia of the left ventricle. In our series, such cardiac defects included interruption of the aortic arch in four infants, and totally or partially anomalous pulmonary venous connection in six.

SELECTION AND MANAGEMENT OF DONORS

Donors and recipients are matched with respect to ABO blood group compatibility. Matching of human leukocyte antigen (HLA) serotypes is not a prerequisite to cardiac transplantation. Evaluation of the donor heart consists of an electrocardiogram to detect any changes in the T wave or ST segments, and echocardiographic evaluation to confirm the shortening fraction of the left ventricle at greater than 25%, and to rule out significant structural abnormalities such as severe mitral regurgitation. Prolonged cardiac arrest, severe metabolic acidosis, and the need for inotropic support of the donor, are not absolute contraindications to donation, and have not been associated with poor outcome. Of grafts used in our series, approximately one-third had a "down time" ranging from 2 to 60 minutes. Donor hearts have often been inadequately loaded for volume because of diabetes insipidus. When initial echocardiographic function is poor, appropriate replenishment of volume may be necessary, and echocardiography should be repeated after 4 to 6 hours before the potential donor is rejected. Lines to monitor central venous and arterial pressures are established in each donor both for resuscitation and monitoring during procurement of multiple organs.

A degree of mismatch between the sizes of donor and recipient hearts is well tolerated. The average ratio of weights, in our series, was 1.8:1, with a range from 0.6 to 4.6. Undersized donor hearts, nonetheless, can be problematic, particularly in neonates with high pulmonary vascular resistance. A graft from a donor weighing less than four-fifths of the body weight of the recipient may not adequately support the circulation. On the other hand, severely oversized grafts can create a technical challenge, and often require the sternotomy to be left open for several days after transplantation. Few infants who received a significantly oversized graft developed marked left ventricular hypertrophy during the first week after transplantation. This phenomenon, which is not unlike hypertrophic cardiomyopathy, can cause near obliteration of the ventricular cavity, and can be fatal if allowed to progress.

OPERATIVE TECHNIQUE FOR THE DONOR

The technique of procurement of the heart utilises inflow diversion with a single dose, of 250 to 500 mL, of cold, crystalloid cardioplegic solution (Fig. 10.1). This

A

FIGURE 10.1. Removal of the donor heart. A: After dividing the brachiocephalic vein, the branches of the aortic arch are ligated. Division of the duct or ligament exposes the proximal descending the aorta.

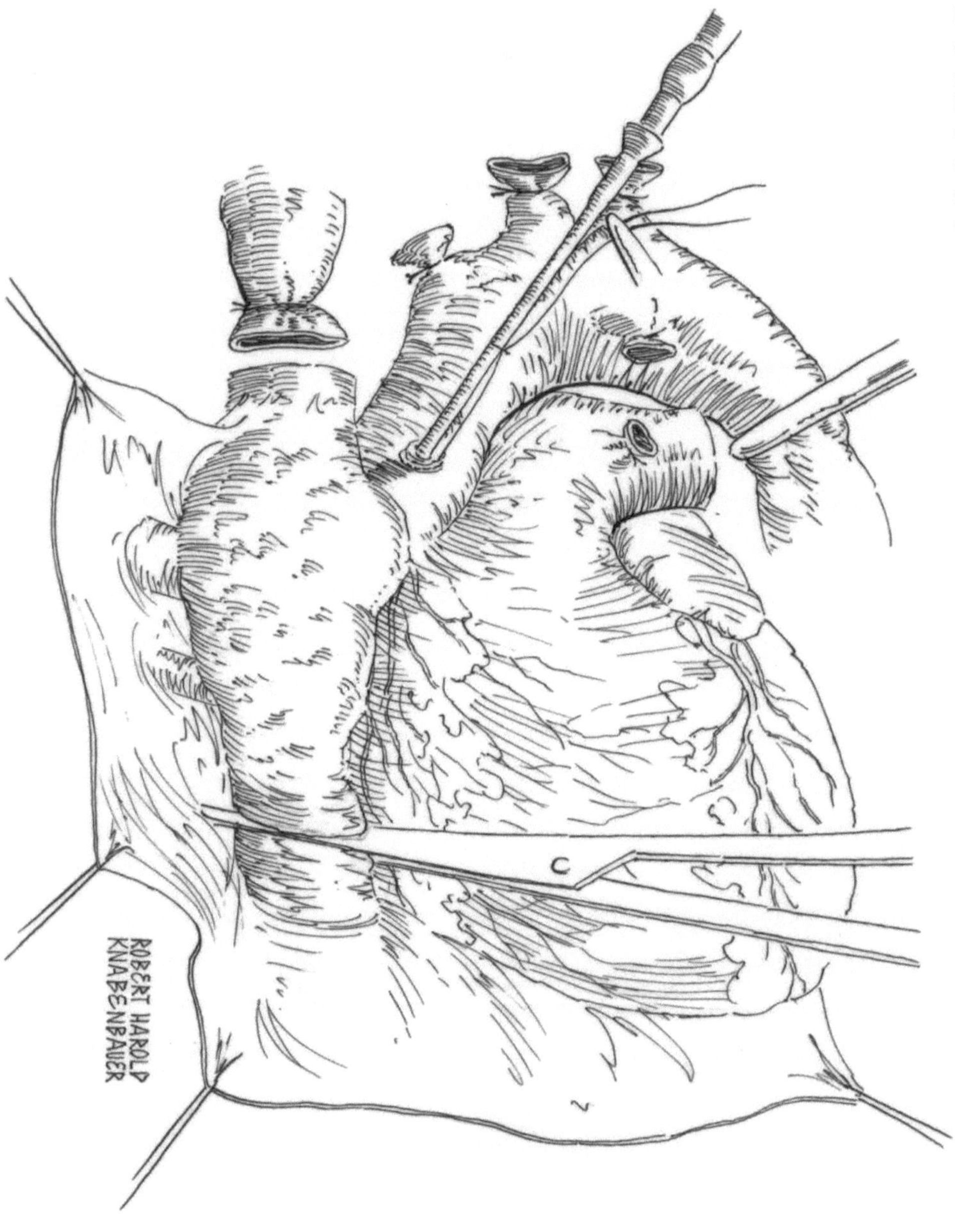

B

Figure 10.1. (*cont.*) B: A catheter for delivery of cardioplegia is positioned in the ascending aorta. The superior caval vein is ligated and divided, and the inferior caval vein is divided. The descending aorta is clamped. The right or left superior pulmonary vein is divided for decompression of the heart.

is followed by removal of the block constituting the heart, great veins, aorta with arch, and the pulmonary trunk. The composition of the intracellular cardioplegic solution includes 27 mEq of sodium chloride, 20 mEq of potassium chloride, 3 mEq of magnesium sulphate, 2.25 mEq of sodium bicarbonate, 250 mq of methylprednisolone, and 1000 mL of 5% dextrose in water. During the operation, the donor is given heparin at a dose of 5 mg/kg, and 50% dextrose solution, given as 2 to 3 mL/kg intravenously. The infusion of dextrose is repeated every 15 minutes until the organ is excised. The graft is placed in a saline bath maintained at 4°C. The bag is placed in a sterile and sealed container, and immersed in an ice chest. The safe limit for ischaemic times is not known in children. In our experience, donor hearts have been successfully implanted despite prolonged ischaemia of up to 9.5 hours. In addition to the thymus, the spleen and the mesenteric lymph nodes are removed from the donor, frozen, and stored for later immunological testing.

OPERATIVE TECHNIQUE FOR THE RECIPIENT

Continuous infusion of cyclosporine at 0.1 mg/kg/h is started in the recipient when a suitable organ is secured. Anaesthetic management should focus on maintaining a balance between the pulmonary and systemic circula-

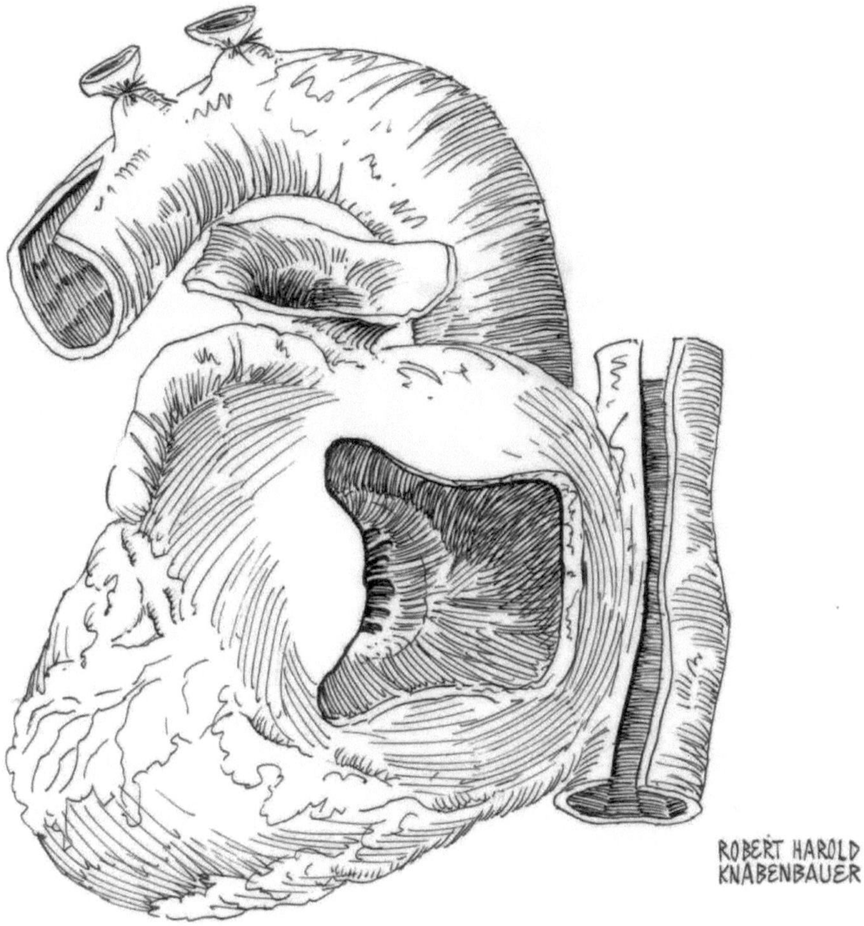

C

FIGURE 10.1. (*cont.*) C: The heart prepared for implantation. The pulmonary venous orifices are connected, the aortic arch is opened along its convex margin, and the right atrium is opened along a posterior incision from the inferior to the superior caval veins, staying close to the interatrial groove to avoid the sinus node.

tions, and avoiding myocardial depressant agents (see also Chapter 9). We administer methylprednisolone, at 25 mg/kg and cephalosporin antibiotic at 25 mg/kg, and surround the head with ice packs. The method of orthotopic cardiac allotransplantation and reconstruction of the aortic arch in infants with hypoplasia of the left ventricle was first described by Bailey and his colleagues.[11] The technique was later modified to reduce the duration of circulatory arrest[12] (Fig. 10.2). The chest is opened through a midsternal incision, and the pericardium is incised vertically. Near-total thymectomy is done. Cardiopulmonary bypass with asanguinous prime is initiated using a single venous drainage cannula in the right atrium, and an arterial perfusion cannula placed in the pulmonary trunk but directed into the duct and snared to prevent flow into the lungs. Systemic cooling to the core rectal temperature of 18° to 20°C lasts from 15 to 25 minutes. During this period, the branches of the aortic arch are dissected. Low-flow perfusion is maintained using an active sucker in the opened right atrium. The diminutive ascending aorta is divided and ligated just proximal to the brachiocephalic artery. Traction on this ligature facilitates better exposure of the proximal descending aorta. The pulmonary trunk is transected at the level of the valve. The hypoplastic heart is excised, leaving in place a cuff of posterior atrial walls and a rim of septum.

After inspecting the pulmonary veins, the implantation of the donor heart begins with the atrial septum at its inferior aspect, using a continuous monofilament suture that is carried around the right atrium. A flexible suction catheter is repositioned in the right atrial appendage while the left atrial anastomosis is completed. With the infant in the Trendelenburg position, the aortic arch vessels are loosely snared, the circulation is arrested, the arterial duct is ligated and divided, and all ductal tissue is excised. The undersurface of the aortic arch is incised from the level of the brachiocephalic trunk to several millimeters beyond the duct onto the descending aorta. The neoaortic arch is reconstructed using the opened long segment of donor arch starting beyond the amputated duct. A period of 15 to 20 minutes of circulatory arrest provides a bloodless field for this part of the procedure. Special

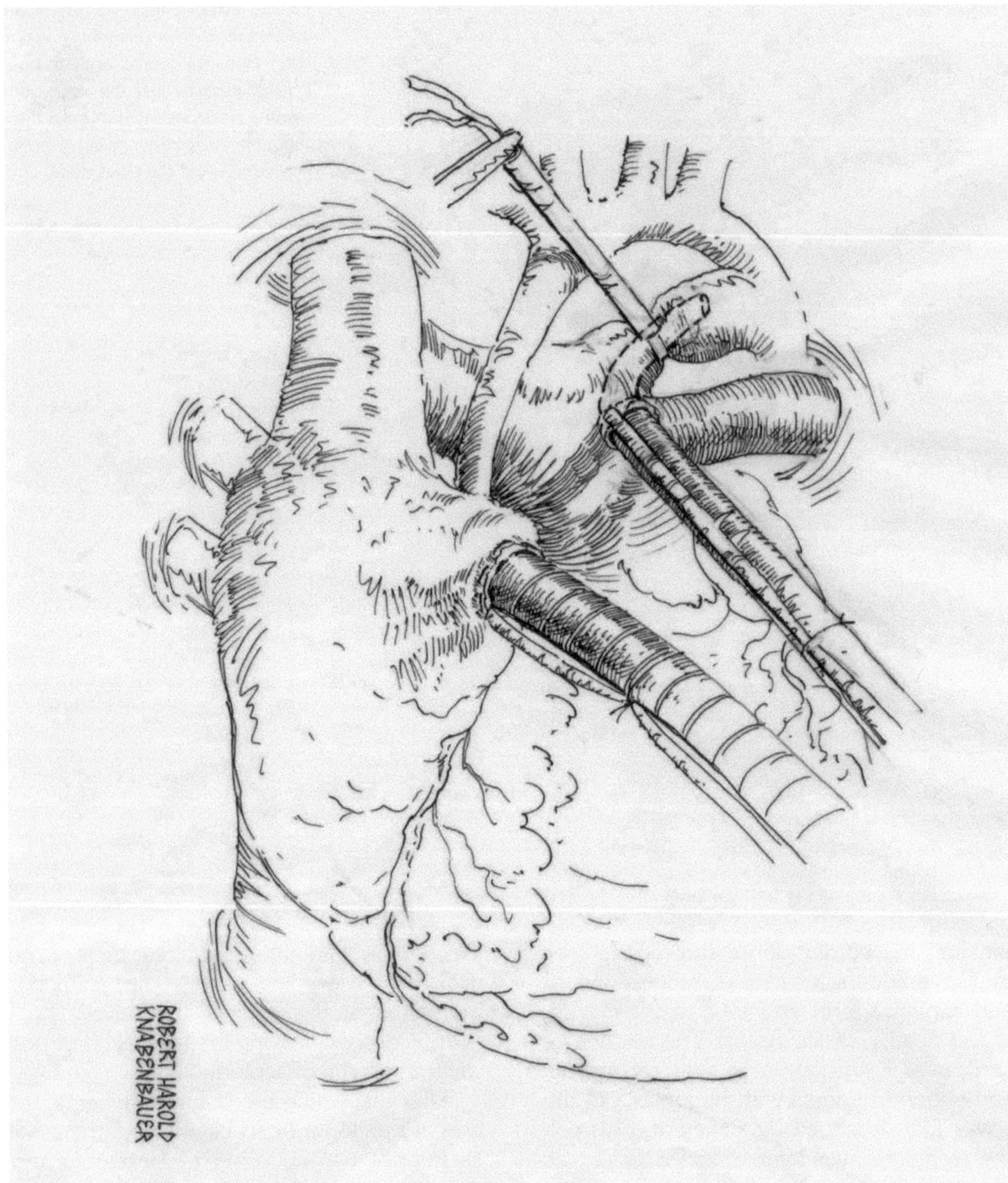

A

Figure 10.2. The technique for transplantation. A: The arterial inflow cannula is introduced in the pulmonary trunk and directed into the duct. A single venous cannula is placed in the right atrial appendage.

care is exercised to identify and avoid injury of the recurrent laryngeal nerve. The aorta is filled with saline at 4°C via the stump of the donor brachiocephalic artery, which is also used for insertion of the arterial cannula. The venous cannula is placed into the donor right atrium through the appendage. Perfusion is resumed, and air is evacuated through a vent site in the donor ascending aorta. The occluders around the arch vessels are removed and the pulmonary arterial anastomosis is completed during the phase of early recirculation. The patient is rewarmed to 37°C and extracorporeal reperfusion is maintained for a minimum of 60 minutes. In some instances, when cold ischaemic time is prolonged, reperfusion is extended

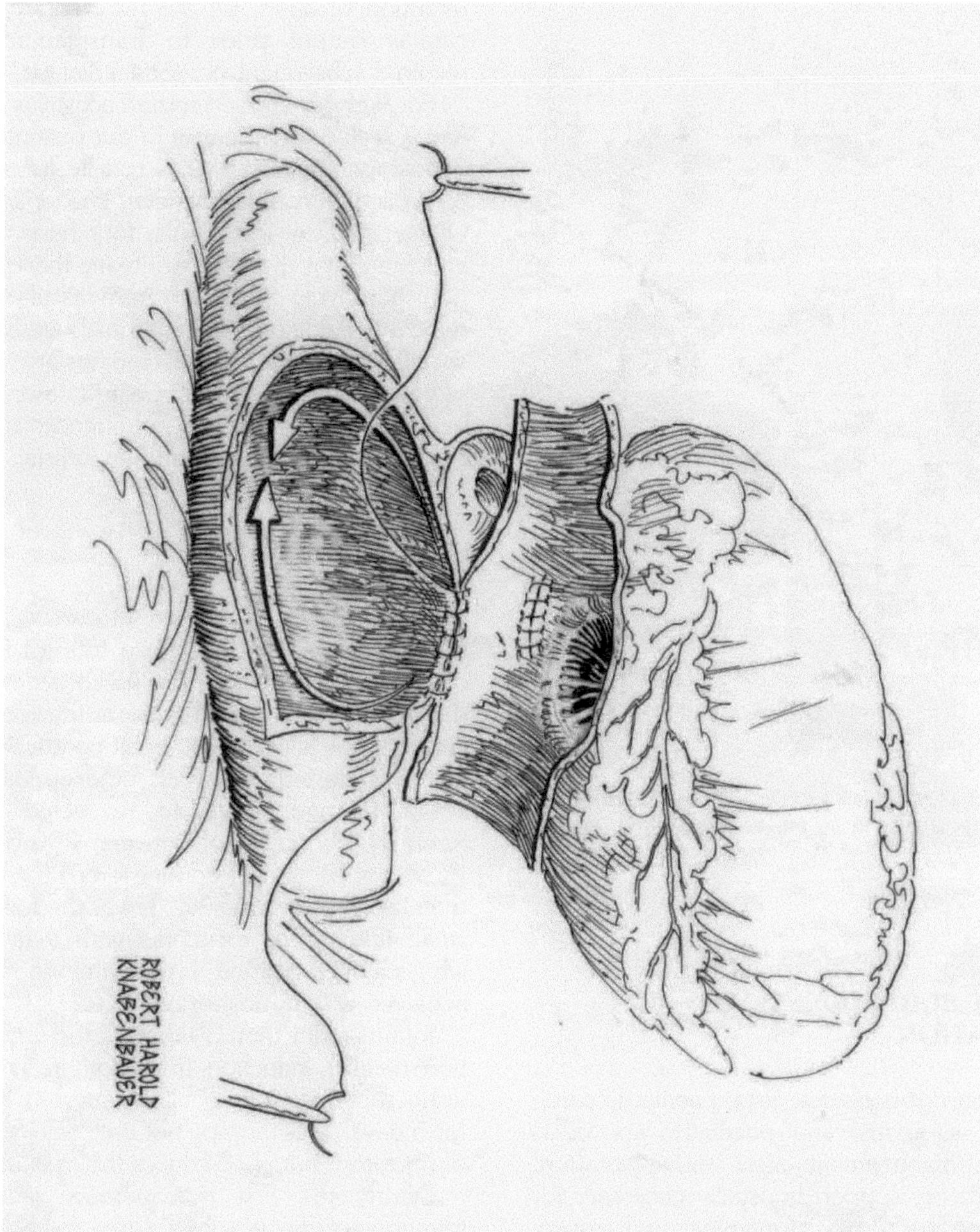

B

Figure 10.2. (*cont.*) B: The branches of the aortic arch are dissected and the pulmonary trunk is transected proximal to the bifurcation. The tiny ascending aorta is divided and ligated proximal to the brachiocephalic trunk. After excision of the heart, low-flow perfusion is maintained with a sucker in the right atrium, and implantation of the donor organ begins with a double-armed suture at the crux. Reconstruction of the septum begins first followed by right atrial anastomosis.

to 80 to 90 minutes to achieve complete functional recovery of the heart. During the latter half of the phase of reperfusion, the electrolytes are normalised and the perfusate is hemoconcentrated to a hematocrit of 28% or higher. Dopamine at 2 to 3 mg/kg/min and isoproterenol at 0.02 to 0.05 mg/kg/min are started prior to separation from bypass. Pacing wires are rarely necessary, as spontaneous cardiac contractions and regular sinus rhythm usually resume by the time the patient has reached normothermia.

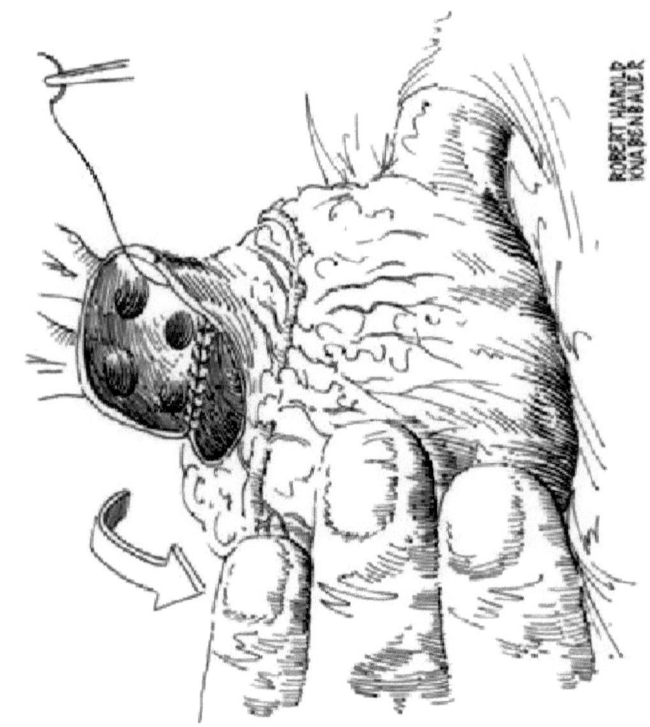

C

FIGURE 10.2. *(cont.)* C: The left atrial anastomosis is performed with the heart tilted up starting at the crux and going in a counterclockwise direction.

MANAGEMENT SUBSEQUENT TO TRANSPLANTATION

A designated group of neonatologists, paediatric cardiologists, cardiac surgeons, and paediatric specialists are involved in management after transplantation. Mechanical ventilatory support is usually necessary for 2 or 3 days. Inotropic drugs, dopamine, and isoproterenol, in low doses, are continued for 3 to 4 days. Infusion of prostaglandin E_1, which has vasodilatory and immunosuppressive properties, is also continued at 0.05 μg/kg/min, and tapered off over a period of 1 week. For recipients of oversized hearts who develop hypertension, the use of chronotropic and inotropic drugs is avoided. Instead, vasodilators and calcium-channel blockers, and rarely beta-blockers, are added. Perioperative acute renal dysfunction is manifested by oliguria and retention of fluid. A rise in blood urea nitrogen is commonly seen. In the majority of cases, this is a self-limited phenomenon that resolves within a few days. Perioperative peritoneal dialysis is safe, and extremely effective in controlling hyperkalemia and retention of fluids. Of our recipients who were in low cardiac output prior to transplantation, 22 (15%) required subsequent peritoneal dialysis.

Focal or generalised seizure activity is not uncommon, and is seen in one-quarter of our patients during the first 3 postoperative days. It is usually transient, and easily controlled with phenobarbital, lorazepam, or diazepam. Of our surviving recipients, four have required chronic anticonvulsant therapy. Antibiotic therapy is maintained perioperatively until all intravascular catheters are removed. Additional antimicrobials are only used to treat specific infections. Enteral feedings are resumed as soon as bowel function returns. Gastrointestinal complications have been rare since we commenced the routine intravenous use of H_2 receptor antagonists.

IMMUNOSUPPRESSION

Methylprednisolone is given intravenously ever 12 hours at a dose of 20 mg/kg for a total of four doses postoperatively. Afterward, the use of steroids is limited to the treatment of acute rejection. Intravenous infusion of cyclosporine is continued postoperatively until oral feedings are started, at which time cyclosporine is given orally. Dosage is adjusted to achieve target trough levels in the serum of between 200 and 300 ng/dL. Azathioprine is also administered, at 3 mg/kg/d, and then tapered to 1 mg/kg/d over the following year. In a small number of recipients with a favourable history who survived beyond 1 year, immunosuppression was achieved with cyclosporine alone.

Infants older than 30 days at the time of transplantation receive induction immunotherapy postoperatively using rabbit thymoglobulin (Sangstat) at 1.5 mg/kg/d for 5 days. This therapy not only delays the first episode of rejection, but also reduces the frequency of rejection.

During the first postoperative week, intravenous immunoglobulin is administered for a total of three to four doses at 400 mg/kg/d. This, in conjunction with prophylactic ganciclovir given for a period of 2 weeks, may reduce the risk of systemic infection with cytomegalorvirus.

SURVEILLANCE

Postoperatively, infants are examined daily for evidence of fever, tachypnea, irritability, poor feeding, unexplained rise in heart rate, arrhythmias, hepatomegaly, and the development of gallop or a new heart murmur. Along with these clinical findings, other noninvasive surveillance techniques are employed to monitor the

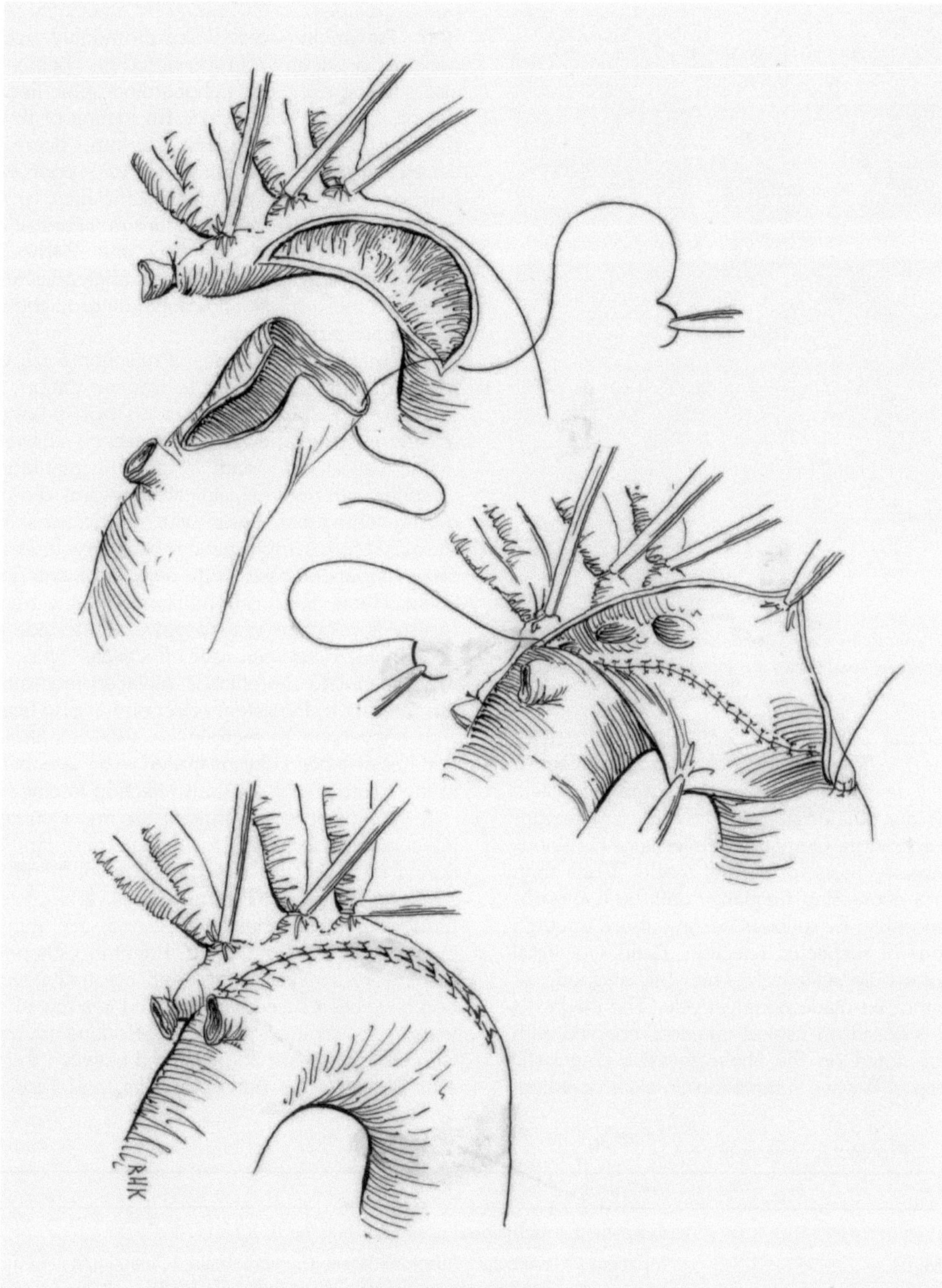

D

Figure 10.2. (*cont.*) D: The snares around the aortic arch vessels are tightened loosely and the circulation is arrested. After ligating the duct, ductal tissue on the aortic side is excised and the reconstruction of the aortic arch starts on the descending aorta with a wide anastomosis.

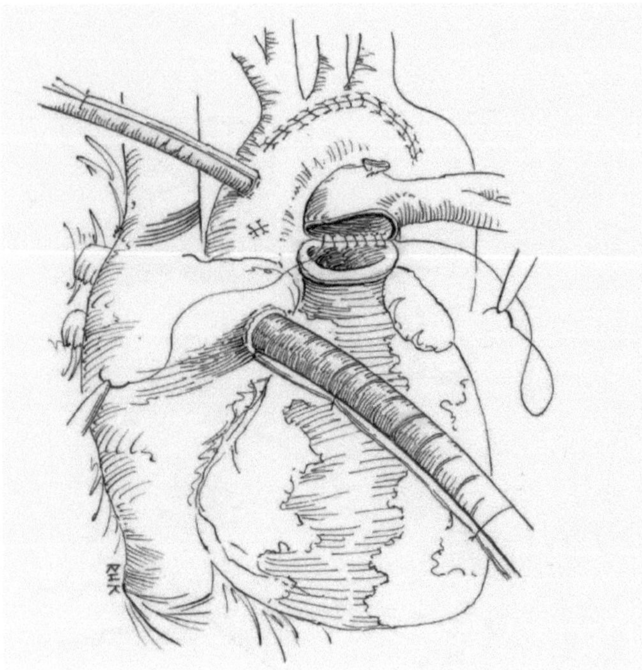

E

Figure 10.2. *(cont.)* E: Cardiopulmonary bypass is reinstituted. The pulmonary arterial anastomosis is completed during reperfusion.

infants in general and the transplanted heart in particular. Serial chest radiographs, echocardiography, electrocardiography, levels of cyclosporine, and complete blood counts are obtained twice weekly during perioperative hospitalisation. Outpatient surveillance continues on a twice-weekly basis for several weeks. Visits then continue on a decreasing frequency until by 6 months they occur monthly. Evaluations become more frequent during periods of suspected rejection. Endomyocardial biopsy is done only selectively when the diagnosis of rejection cannot be made noninvasively. The diagnosis of rejection is based on clinical intuition, coupled with new changes noted on the above invasive diagnostic panel. A drop in R-wave summation on electrocardiography greater than 20% may be associated with rejection. Advancing global cardiomegaly, pulmonary oedema, or pleural effusion noted on chest x-rays may also suggest rejection. Echocardiographic findings consistent with rejection include thickening of the left ventricular posterior wall and septum, decreasing left ventricular fibre shortening fraction, poor ventricular function, new mitral valvular insufficiency, or new pericardial effusion. Recipients undergo a cardiac catheterisation, coronary angiogram, and 24-hour Holter monitoring annually. Coronary angiography may be indicated more frequently, depending on the condition of the coronary arteries.

The majority of episodes of rejection occur within the first 3 months subsequent to transplantation. Although up to one-quarter of infants do not experience any rejection, acute and chronic rejection continue to be major causes of death after transplantation. Mild episodes with minimal clinical symptoms can be treated on an outpatient basis with intravenous pulses of steroids. More symptomatic rejection requires treatment on an inpatient basis with polyclonal anti–T-cell antibodies (Table 10.2). Methotrexate for the treatment of recurrent rejection unresponsive to steroids and antibody therapy has been used effectively.[13] More recently, mycophenolate mofetil has replaced methotrexate for this indication. Persistent rejection has also been treated with conversion to tacrolimus. Total lymphoid irradiation has also been demonstrated to be safe and effective in the treatment of resistant rejection in children.[14]

Immunosuppressed infants are most susceptible to infection immediately following transplantation and during treatment for rejection. Careful monitoring with surveillance, bacterial, viral, fungal, and protozoal cultures, along with clinical vigilance, can minimise the incidence of lethal infection. Pneumocystis pneumonia, respiratory syncytial virus, and cytomegalovirus infection have been successfully treated in a few of our recipients with minimal morbidity. Routine immunizations with killed virus are administered between the third and sixth month after transplantation, and have generally

Table 10.2. Protocol for rejection

Type	*Treatment*	
Mild	Outpatient:	methylprednisolone 20mg/kg IV q12h × 4 d
Moderate–severe	Inpatient:	methylprednisolone bolus; thymoglobulin 1.5mg/kg/d (7–10 d)
	or	Antithymocyte globulin 15mg/kg/d (7–10 d)
	or	Conversion to tacrolimus
	or	Conversion to mycophenolate mofetil
	or	Intravenous immunoglobulin at 2g/kg
Recalcitrant–chronic		Total lymphoid irradiation
Irreversible with cardiovascular collapse		Extra corporeal membrane oxygenation

produced an adequate host response. The incidence of childhood infections, such as otitis media, gastroenteritis, or upper respiratory infections, is the same as for children in general.

GROWTH AND DEVELOPMENT

Cognitive development in infants subsequent to cardiac transplantation falls within the average range. It is important to note that the question of underlying brain abnormalities has been raised in children with hypoplasia of the left ventricle.[15] At Loma Linda University, a preliminary study of infants receiving transplanted hearts used the Bayley scales of infant development to test mental and psychomotor development and to examine infant behaviour.[16] The mean developmental indexes for both mental skills, at a mean of 87, and motor skills, at a mean of 90, were within normal limits. Infants undergoing prolonged periods of hypothermic circulatory arrest performed significantly worse than those with shorter times. In many children, visual motor skills were delayed. These neurodevelopmental findings are not dissimilar from those in other populations of children with chronic medical conditions. Our data concerning early developmental outcome are similar to results for infants who have undergone staged surgical repair for hypoplasia of the left ventricle.[17,18] It is essential, therefore, that the neurological and developmental state be evaluated before and after transplantation by a paediatric neurologist, paediatric occupational therapist, or paediatric specialist (see also Chapter 11). Transient perioperative seizures and hypotonia were the most common neurological abnormalities in our population. Of our recipients, 14 have developed serious neurological sequels, six with cerebral palsy, five with cerebral vascular accidents, three with autism, one of these also having cerebral palsy, and one with a generalized developmental delay. Psychostimulant medication for attention-deficit hyperactivity disorder is being used in four patients, with good results. The very few patients who required prolonged use of corticosteroids for control of recurrent rejection have shown some retardation of growth. This pattern normalised with the cessation of steroids. The majority of our patients have followed normal growth curves, and currently enjoy a reasonable good quality of life.[17]

RESULTS

Data from the registry of the International Society for Heart and Lung Transplantation from the years 1991 to 2000 shows that, on average, 92 hearts are transplanted yearly into infants less than 1 year of age.[19] Complex congenital cardiac disease is the predominate indication. The 30-day mortality in this group is 20%, with actuarial survival at 1 and 2 years of 69% and 65%, respectively. The operative mortality of transplantation for hypoplasia of the left heart in our population was 11%. Acute rejection, along with technical and management failures, accounted for more than half of our early deaths. The encouraging intermediate-term results (Fig. 10.3), with actuarial survival of 84%, 73%, 67%, and 61% at 1, 5, 10, and 15 years, respectively, warrant continued use of transplantation.

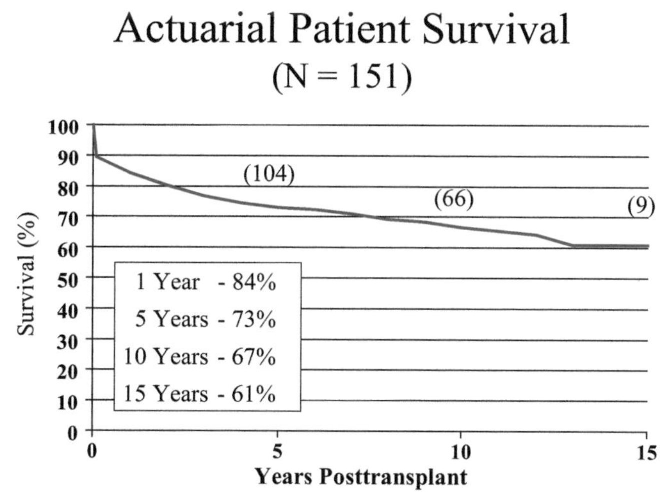

FIGURE 10.3. Actuarial survival of infants with hypoplasia of the left ventricle following cardiac transplantation.

CURRENT ISSUES AND FUTURE PERSPECTIVES

Chronic immunosuppressive therapy in infants and children has been associated with renal dysfunction and hypertension. Of our survivors, 26 (17%) have required chronic antihypertensive medications for 1 year or more. Of those who survived more than 2 years and for whom data are available, one-third have a glomerular filtration rate of less than 60 mL/min/1.73 m^2. At the present time, two long-term survivors, 12 and 14 years after transplantation, are suffering from chronic renal failure and are awaiting renal transplantation. Renal impairment is related to cyclosporine toxicity, as well as a preoperative state of low perfusion and damage initiated by extracorporeal circulation. When we treat deterioration of renal function, we have had good success with verapamil or diltiazem. Both of these calcium-channel blockers compete with cyclosporine for metabolism, allowing lower cyclosporine dosing. In addition, they counteract the vasoconstrictive properties of cyclosporine.[17]

Life-threatening neoplasms, particularly lymphomas, are a well-known complication in recipients of transplants. The occurrence of these tumours may be related to the more intense immunosuppressive therapy that is sometimes given to reverse rejection. The incidence of lymphoproliferative tumours among recipients of cardiac allografts has varied from 2.3% to 10%.[20] In our series, 8 out of 151 (5.3%) have developed lymphoproliferative disease subsequent to transplantation (Fig. 10.4) at a median of 7.6 years after the procedure, with a range from 1.6 to 12.5 years. We believe that avoiding the chronic use of steroids and maintaining a level of immunosuppressive therapy as low as is compatible with good allograft function reduces the incidence of neoplasms and hypertension. The search for better immunosuppressive agents must continue. An ideal drug would induce allograft-specific unresponsiveness and prevent cellular and humoral rejection, but maintain normal non–allograft-related immunity with minimal toxicity. Another solution to this problem lies in the promise of induction in the host of donor-specific immunological tolerance. When such tolerance becomes a clinical reality, recipients may no longer require lifelong immunosuppressive treatment.

The major deterrent to long-term survival of recipients of transplanted hearts is development of coronary arterial disease in the transplanted heart. The pathobiology and natural history of this lesion in children is not yet well understood. The classic feature, however, is concentric, diffuse, myointimal hyperplasia leading to luminal narrowing and obliteration of the major epicardial vessels and branches along their entire length. The etiology is multifactorial. Factors that may contribute to vasculopathy include complex immune-mediated interactions between donor endothelial cells and host mononuclear cells, immunosuppressive drugs, ischaemic or reperfusion endothelial injury, cytomegaloviral infection, and late severe rejection.[21] Coronary angiography has been the technique of choice for surveillance. This technique, however, tends to underestimate the severity of the disease. Exercise echocardiography and intracoronary ultrasonography can aid in making earlier diagnosis of this serious complication.[22,23]

In our population (Fig. 10.5), the incidence of coronary arterial disease subsequent to transplantation is 11.3%, being seen in 17 out of 151 patients. The median time to the development of coronary arterial disease in our series has been 6.3 years, with a range of 1.9 to 14.5 years. It is unclear whether the use of maintenance steroids is an independent risk factor for the development of coronary arterial disease, or whether it is the repeated rejections forcing the use of steroids that is the culprit. It is our impression that control of rejection by effective steroid-free immunomodulation may perhaps lower the incidence of coronary arterial disease. The customary treatment for graft vasculopathy is retransplantation. In our series (Fig. 10.6), 12 patients (8%) have undergone retransplantation, all of whom have survived. Of these 12 patients, 9 had coronary arterial disease as their primary indication.[24]

Cardiac transplantation in infants, with all its limitations, has proven to be an effective treatment for the most severe forms of congenital cardiac disease. The search for more sensitive techniques of detecting early

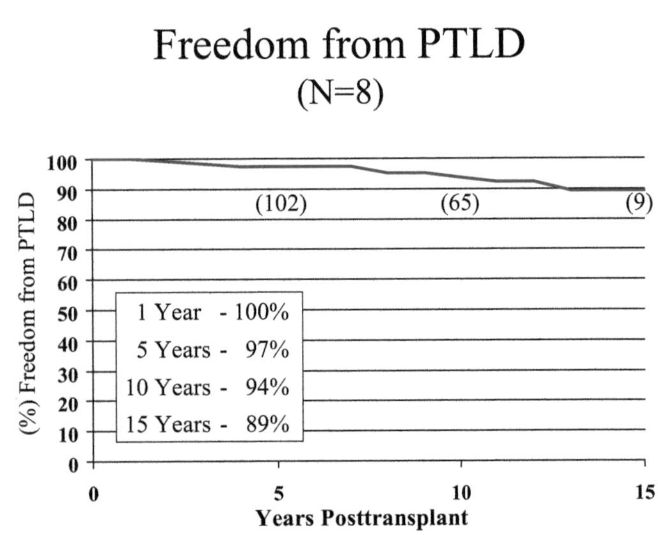

FIGURE 10.4. Actuarial freedom from lymphoproliferative disease subsequent to transplantation.

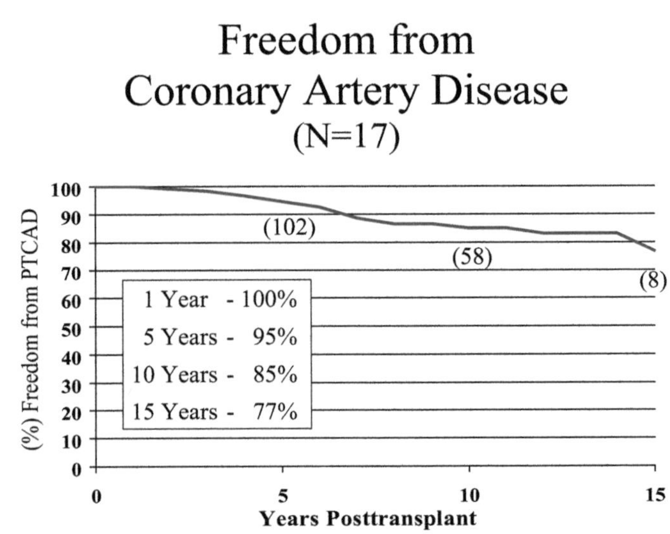

FIGURE 10.5. Actuarial freedom from coronary arterial disease subsequent to transplantation.

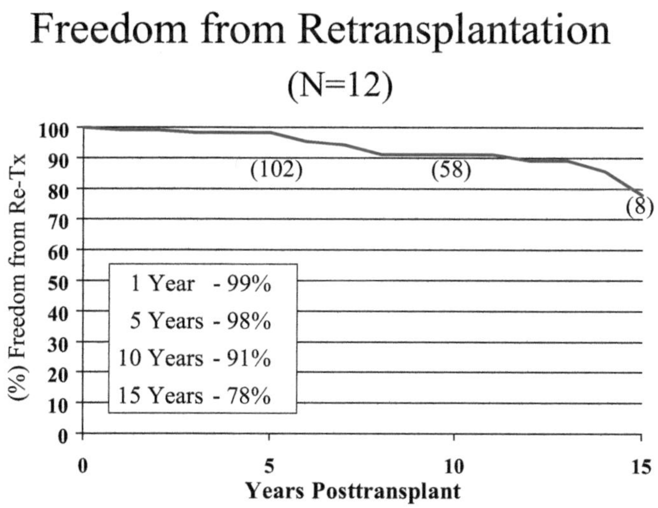

FIGURE 10.6. Actuarial freedom from retransplantation.

rejection continues. Long-term survival will be possible as immunobiology induces acquired tolerance and achieves more selective and effective control of the immune system. Such advances, along with increased availability of donor hearts, can make transplantation a more practical and durable therapy for infants with hypoplasia of the left heart.

References

1. Norwood WI, Lang P, Hansen DD. Physiologic repair of aortic atresia—hypoplastic left heart syndrome. N Engl J Med 1983;308:23–26.
2. Iannettoni MD, Bove EL, Mosca RS, et al. Improving results with first-stage palliation for hypoplastic left heart syndrome. J Thorac Cardiovasc Surg 1994;107:934–940.
3. Mahle WT, Spray TL, Wernovsky G, Gaynor JW, Clark BJ 3rd. Survival after reconstructive surgery for hypoplastic left heart syndrome: a 15-year experience from a single institution. Circulation 2000;102(19 suppl 3):III136–141.
4. Bailey LL, Nehlsen-Cannarella SL, Doroshow RW, et al. Cardiac allotransplantation in newborns as therapy for hypoplastic left heart syndrome. N Engl J Med 1986;315:949–951.
5. Chiavarelli M, Gundry SR, Razzouk AJ, et al. Cardiac transplantation for infants with hypoplastic left heart syndrome. JAMA 1993;270:2944–2947.
6. Razzouk AJ, Chinnock RE, Gundry SR, et al. Transplantation as a primary treatment for hypoplastic left heart syndrome: intermediate-term results. Ann Thorac Surg 1996;62:1–8.
7. Bailey LL, Nehlsen-Cannarella SL, Concepcion W, Jolley WB. Baboon-to-human cardiac xenotransplantation in a neonate. JAMA 1985;254:3321–3329.
8. Johnston J, Sakala EP. Neonatal cardiac allotransplantation facilitated by in-utero diagnosis of hypoplastic left-sided heart syndrome. West J Med 1990;152: 70–72.
9. Chinnock RE, Peverini RL. Beginnings and endings. Pediatr Transplant 1999;3:92–94.
10. Ruiz CE, Gamra H, Zhang HP, et al. Stenting of the ductus arteriosus as a bridge to cardiac transplantation in infants with hypoplastic left heart syndrome. N Engl J Med 1993;328:1605–1608.
11. Bailey LL, Concepcion W, Shattuck H, et al. Method of heart transplantation for treatment of hypoplastic left heart syndrome. J Thorac Cardiovasc Surg 1986;92:1–5.
12. Vricella LA, Razzouk AJ, del Rio M, Gundry SR, Bailey LL. Heart transplantation for hypoplastic left heart syndrome: modified technique for reducing circulatory arrest time. J Heart Lung Transplant 1998;17: 1167–1171.
13. Chinnock RE, Larsen RL, Emery JR, et al. Pretransplant risk factors and cause of death or graft loss at heart transplantation during early infancy. Pediatric Heart Transplant Team, Loma Linda. Circulation, 1995; 92:II206–209.
14. Asano M, Gundry SR, Razzouk AJ, et al. Pediatric Heart Transplantation Group. Total lymphoid irradiation for refractory rejection in pediatric heart transplantation. Ann Thorac Surg 2002;74(6):1979–1985.
15. Glauser TA, Rorke LB, Weinberg PM, et al. Congenital brain anomalies associated with the hypoplastic left heart syndrome. Paediatrics 1990;85:984–990.
16. Baum M, Freier MC, Freeman KR, Chinnock RE. Developmental outcomes and cognitive functioning in infant and child heart transplant recipients. Prog Pediatr Cardiol 2000;11:159–163.
17. Chinnock RE, Cutler D, Baum M. Clinical outcome 10 years after infant heart transplantation. Prog Pediatr Cardiol 2000;11:165–169.
18. Rogers BT, Msall ME, Buck GN, et al. Neurodevelopment outcome of infants with hypoplastic left heart syndrome. J Paediatr 1995;126:496–498.
19. Boucek MM, Edwards LB, Keck BM, et al. The Registry of the International Society for Heart and Lung Transplantation: Fifth Official Pediatric Report—2001 to 2002. J Heart Lung Transplant 2002;21:827–840.
20. Boyle GJ, Michaels MG, Webber SA, et al. Posttransplantation lymphoproliferative disorders in pediatric thoracic organ recipients. J Pediatr 1997; 131(2):309–313.
21. Mulla NF, Johnston JK, Vander Dussen L, et al. Late rejection is a predictor of transplant coronary artery disease in children. J Am Coll Cardiol 2001;27:243–250.
22. Larsen RL, Applegate PM, Dyar DA, et al. Dobutamine stress echocardiography for assessing coronary artery disease after transplantation in children. J Am Coll Cardiol 1998;32:515–520.

23. Dent CL, Canter CE, Hirsch R, Balzer DT. Transplant coronary artery disease in pediatric heart transplant recipients. J Heart Lung Transplant 2000;19:240–248.
24. Dearani JA, Razzouk AJ, Gundry SR, et al. Pediatric cardiac retransplantation: intermediate-term results. Ann Thorac Surg 2001;71:66–70.

Chapter 11

The Nursing Management of Neonates and Infants with Hypoplasia of the Left Heart

Janet L. Taylor, Fiona Horrox, Sharon Goomany, Sue Smith and Akhlaque N. Bhat

The successful care of infants with hypoplasia of the left heart is very much dependent on a multidisciplinary approach. The nursing management of these infants is particular and precise, so it may also be argued that a team of dedicated nurses with specialist cardiac knowledge is required.

The management begins with diagnosis, which may have been made antenatally. Antenatal diagnosis enables the parents to acquire information about the options available for their baby, and allows them to make an informed decision about their future care. There are three options available to the family: the Norwood sequence, cardiac transplantation, and no intervention. In the latter instance, the infant is made comfortable and allowed to die peacefully. This chapter discusses the management of babies whose families opt for the Norwood procedure. This is the option for treatment employed by our unit and by the majority of cardiac centres throughout the world that look after this very sick group of babies. We also briefly discuss the advantages from a nursing point of view of using a shunt placed directly from the right ventricle to the pulmonary arteries rather than the classical Blalock-Taussig shunt, a modification introduced by Sano and his colleagues[1] and that has been used routinely by our surgical team over the last 2 years.

TIMING OF DIAGNOSIS

If the diagnosis is made antenatally, the mother should be booked into a maternity unit that has specialist facilities for neonatal intensive care and that is close to a paediatric cardiac centre. Diagnosis at this stage allows the parents to prepare and to plan for a short separation from their baby, and to gather information and knowledge. They will be able to speak to the cardiac liaison staff and to visit the cardiac unit, including the intensive care unit.

If, however, the diagnosis is not made until after the infant is born, which is the more usual situation, plans must be made for immediate transfer to a suitable centre by a specialist transport team. Before the journey, the infant must be stabilised at the referring hospital to minimise the risks of transfer. One of the overall preoperative principles is to establish whether the other major organs are normal. Of this group of children, one-tenth will have other anomalies of other major organs.[2] Congenital malformations involving the nervous system have been shown to exist in almost one-third of babies with hypoplasia of the left heart. The infants may also develop problems with the central nervous system related to the ischaemic, hypoxic events that can occur in their unstable state. Events that may affect the central nervous system include hypoxia, acidosis, hypercarbia, hypotension, coagulopathies, and hypo- or hyperglycaemia.[3] Clinical assessment of the brain, kidneys, liver, and gut take place prior to surgery. This includes clinical tests for full blood count, urea, electrolytes, clotting profile, and liver function. The nurse assists in the collection of blood for these tests, and coordinates other diagnostic investigations, such as cerebral and renal ultrasonic scans. It is important to send blood for chromosomal analysis at this point, prior to administration of any blood products. A major insult to, or abnormality of, other organs increases the chances of morbidity and mortality.

Before admitting the baby to the intensive care unit, the nursing staff need to collate as much information as possible, for example, name, date of birth, weight, presumed diagnosis, and the present condition of the baby. This allows the bed space to be set up safely with equipment that is appropriate to the baby's age and weight. It is important to ensure that all equipment is in full working order. When considering the need for emergency oxygen and suction, it is important to remember that an air/oxygen blender should be available. This allows the infant to be hand-ventilated in a concentration of oxygen of less than 100%, and also in air.

The main principle of nursing management is to ensure that the fine equilibrium between the systemic and pulmonary circulation is maintained, both preoperatively and postoperatively. This chapter discusses management using a systems approach. It is useful to remember that some of the nursing interventions can be applied to those with other complex congenital cardiac malformations. The continuous assessment and observation of vital signs and reviewing the results from laboratory tests are two of the most important roles undertaken by the nurses in the intensive care unit. Subtle changes in the measured parameters may have a great impact on the condition of the patient and should be reported immediately to the medical staff. Constant observation and awareness will alert the nurse quickly to such changes.

THE RESPIRATORY SYSTEM

The general principles of respiratory management remain similar both preoperatively and postoperatively. Blood pH and levels of carbon dioxide and oxygen are manipulated cautiously to maintain the fine balance between systemic and pulmonary vascular resistance.

Not all infants require mechanical ventilation. This depends on variables such as the haemodynamic state and the amount of prostaglandin required to maintain ductal patency. Many infants can be successfully nursed on paediatric cardiac wards prior to surgery. All of these infants, nonetheless, require ventilation in the postoperative period, and for varying lengths of time. For those infants who need to be ventilated, the safety of the endotracheal tube is the most important consideration. Nasal or oral endotracheal tubes should be well secured with tape to prevent accidental extubation. Nasal tubes are preferred for stability and ease of nursing management. A chest x-ray is required to check the position of the tube.

It is important regularly to assess and evaluate arterial blood gases, as small changes in either the systemic or pulmonary vascular resistances, or the lung compliance, may have a big impact on ventilatory requirements. It is preferable to make small changes with regular monitoring, rather than making large changes that may tip the balance. It is especially important to keep a close eye on the respiratory parameters when the infant is weaning and beginning to take over its own respiratory function. At this stage, inspired oxygen should be kept low in order to maintain the partial pressure of oxygen within a range of 35 to 45 mm Hg. Oxygen is a potent pulmonary vasodilator, and it increases the flow of blood through the lungs if given in excessive amounts. As the saturation of oxygen rises, more blood enters the lungs at the expense of the systemic circulation, therefore giving rise to a higher saturation and a lower cardiac output.

Most infants with hypoplasia of the left heart have increased pulmonary blood flow, both in preoperative and postoperative settings, because the flow of blood from the arterial duct in the preoperative period, or from the systemic-to-pulmonary arterial shunt in the postoperative period, preferentially enters the low resistance circuit of the pulmonary circulation. Excess pulmonary blood flow causes systemic hypotension, metabolic acidosis, poor peripheral perfusion, and pulmonary oedema. Signs of poor systemic perfusion, such as widening of the gap between core and peripheral temperatures, declining urinary output, and a generally mottled appearance, should be looked for and immediately reported. The nurse should assess the consistency, colour, and amount of endotracheal secretions in order to detect pulmonary oedema. Interventions to optimize the amount of blood flowing to the lungs include maintaining pH on the acidotic side, at around 7.35, increasing the peak end-expiratory pressure to increase mean airway pressures, and decreasing systemic arterial resistance to promote systemic perfusion. Although some studies indicate the need to add inspired carbon dioxide,[4] the control and manipulation of the ventilatory parameters has proved to be vital in the control of pulmonary vascular resistance. In a minority of patients presenting with increased pulmonary vascular resistance and a low saturation of oxygen, the inspired oxygen may need to be increased and ventilation manipulated to achieve a pH of 7.5 to 7.6. This should improve the flow of blood to the lungs.

In the ventilated patient, suctioning the endotracheal tube is an important part of clearing secretions and maintaining tubal patency. Suctioning the endotracheal tube is a stressful stimulus to the patient, and should be undertaken only following a respiratory assessment. If infants require suction, it is important to make sure that they are adequately sedated. Fentanyl given intravenously as a continuous infusion ensures adequate sedation and analgesia, and minimises an increase in pulmonary vascular resistance by dulling the stress response to noxious stimuli.[5,6] Ideally an in-line system should be used for suction to prevent disconnection from the ventilator and the need for hand ventilation. On the other hand, this system increases the dead space and may make the baby unstable. Shallow suctioning must be employed, using an appropriately sized catheter to reduce the incidence of right upper lobe collapse and hypoxia.[7] The suction pressure should be set at a maximum of 20 kilopascals. Severe hypoxia will upset the balance between pulmonary and systemic vascular resistance, and therefore should be avoided during suctioning.[7,8] Secretions should be assessed for the presence of pulmonary oedema, infection, haemorrhage, and the amount aspirated. Appropriate humidification of the inspired gases helps to reduce the need for excessive physiotherapy and suctioning by preventing the secretions from becoming thick and sticky.

Postoperatively, infants who develop a high pulmonary vascular resistance may require nitric oxide to be added to the inspired gases. Nitric oxide is a selective vasodilator that relaxes the pulmonary vasculature and increases pulmonary blood flow. Nitric oxide is added to the inspiratory side of the ventilation circuit, with a sampling device to monitor levels. It is measured in parts per million, and is recorded hourly alongside the other ventilatory parameters. It is delivered in the concentration of up to around 20 to 25 parts per million. In the expiratory side of the circuit, the level of nitrogen dioxide, which is a waste gas, is measured and recorded. Because of its pulmonary dilatory effects, it is important that it is delivered in a well-ventilated area with a scavenging system. Environmental monitors are also needed, to give an early warning of any leak from the cylinders or circuit. The system for hand ventilating the patient must also be adapted to take the flow of nitric oxide.

Nitric oxide is not without its risks, which include methaemoglobinaemia[9,10] and prolonged bleeding due to suppression of aggregation and adhesion of platelets. This may increase the risk of intraventricular haemorrhage in premature infants and neonates.[11] Nitric oxide may also initially cause a drop in cardiac output due to pulmonary steal as more blood flows to the lungs at the expense of the systemic circulation. The long-term effects are currently unknown but are thought to accelerate the systemic inflammatory response.

To ensure that the infant is comfortable, pain-free, and compliant with the ventilator, it is important to administer appropriate sedation and analgesia. In these infants, as previously stated, fentanyl is often the drug of choice. Midazolam and a paralysing agent may also be added to achieve compliance. Adequate sedation and paralysis will reduce the lability of pulmonary vascular resistance.[6] It is important that the nurse makes a regular assessment of the level of sedation and comfort. This may involve scoring the pain. In the child who is heavily sedated and paralysed, it is important to check pupillary reactions and changes in heart rate and blood pressure in response to stimulation, in order to assess the level of consciousness and comfort. It is very important to keep the infant free from stress. Extra doses of analgesia may be given prior to any nursing procedure. It is important to try to organise combined procedures so as to minimize the need for handling.

THE CARDIOVASCULAR SYSTEM

Though the general principles of cardiovascular management are the same, some issues are unique to the preoperative and postoperative cardiovascular management of those with hypoplasia of the left heart. Preoperatively, it is vitally important to maintain ductal patency, because in the absence of antegrade flow through the ascending aorta the duct constitutes the only source of blood flow to the systemic circulation. Ductal patency is achieved by administration of prostaglandin E_1 at 5 to 100 ng/kg/min. It is infused intravenously, preferably via a central vein, although it can also be given peripherally. These infants should always be nursed with more than one source of venous access. Peripheral lines should be carefully observed for signs of extravasation and of increasing dermal redness. Bolusing of prostaglandin should be avoided for fear of systemic hypotension, so it is highly desirable to achieve administration as an isolated infusion. The infant must be observed for apnoea, hypotension, hypernatraemia, seizures, hyperglycaemia, fluid retention, cyanosis, flushing, and pyrexia, all of which are recognised side effects of treatment with prostaglandin. Apnoea caused by infusion of prostaglandin is often the cause of preoperative ventilation and admission to the intensive care unit. Even in the presence of an ongoing prostaglandin infusion the baby should be constantly observed for signs of ductal closure. Ductal closure leads to low systemic perfusion and its consequences. Earliest signs include increasing tachycardia, widening of the gap between core and peripheral temperatures, and increasing acidosis. If left untreated, this quickly progresses to frank shock manifested by systemic hypotension, poor peripheral pulses, mottled appearance, oliguria or anuria from poor kidney perfusion, signs of necrotising enterocolitis because of poor gut perfusion, eventual sepsis, and death. The nurse should constantly be on the alert for signs and symptoms that might betray ductal insufficiency. Apart from constant observation of parameters that are displayed in real time, for example, core and central temperatures, pulse rate, blood pressure, central venous pressure, and arterial blood gases, a keen eye should be kept on the trend of urinary output, abdominal girth, and bleeding via the rectum to detect any damage to target organs.

The cardiovascular system of patients after the first stage of the Norwood procedure exists within a fine balance of interactions between the events of the intraoperative period the success of the surgical procedure, and the ability of the organs, especially the heart, to recover from the deleterious effects of cardiopulmonary bypass, myocardial ischaemia, and circulatory arrest. All of these factors have a direct influence on postoperative cardiac output. Because of a period of aortic cross-clamping, the heart is deprived of its vascular supply, which affects its ability to contract in the immediate postoperative period. Cardiac contractility is enhanced by the use of inotropes. Depending on preferences within the unit, dobutamine, dopamine, adrenaline, and enoximone are used either in isolation or in combination. The nurse is responsible for the constant and smooth delivery of inotropes to the patient. It is important not to cause peaks and troughs when changing syringes, so the choice of infusion pump is important. Hourly quantities of the drug given should be recorded as micrograms per kilogram per minute (μg/kg/min), and any changes noted. Bolusing of inotropes should be avoided. Blood pressure is recorded half-hourly, although constant monitoring is maintained with arterial access connected to a transducer. The arterial line site should be regularly inspected for signs of haemorrhage. The limb distal to the site of arterial cannulation should be observed for signs of vascular insufficiency, such as capillary filling, temperature, and the strength of the distal pulse.

In the setting of hypovolaemia, the preload can be manipulated by infusing preparations of colloid or blood, depending on the haematocrit. Care must be taken not to bolus colloids in large amounts, because this can acutely distend the heart, resulting in rapid and severe haemodynamic collapse. In the setting of poor cardiac output, reduction of afterload with vasodilators may be needed to minimise myocardial demand for oxygen, and to promote peripheral perfusion. Reduction of afterload is undertaken cautiously, as it may also promote pulmonary perfusion. If vasodilators are prescribed, the nurse must be prepared to infuse volume expanders, because the resulting hypotension can be quite severe. Prompt detection and aggressive management of arrhythmias is vital for babies with the Norwood physiology. Supraventricular arrhythmias are common, and they can potentially decrease cardiac output. The cause of the arrhythmia should be identified and treated to restore maximum cardiac output. Continuous monitoring of the electrocardiogram, with the facility to run a rhythm strip, is a rule. The nurse should be aware that electrolytic imbalance is often the cause of an arrhythmia. Levels of calcium, potassium, and magnesium should be regularly checked and optimised. A pacing box should be readily available for possible intervention. In the event of a junctional ectopic tachycardia, the baby may need to be cooled to between 34 and 35°C, since this arrhythmia is generally resistant to pharmacological measures. A cooling blanket easily achieves these temperatures. It is very important to keep the temperature within this fine range, because uncontrolled cooling to temperatures below 30°C can precipitate ventricular fibrillation. Patients who need to be cooled actively should be routinely sedated and paralysed, because in an unparalysed patient shivering can raise the core temperature to dangerously high levels.

Recently, our surgical team has adopted the technique of using a conduit from the right ventricle to the pulmonary arteries as the source of pulmonary blood flow, rather than the conventional systemic-to-pulmonary arterial shunt. With this technique, the right ventricle supports the pulmonary circulation. The coronary and systemic circulations are not in competition with the pulmonary circulation, and as such are not influenced by changes in pulmonary vascular resistance. The saturations are usually maintained in the 90s. Unlike the conventional systemic-to-pulmonary arterial shunt, regular fine-tuning of ventilatory parameters is not required to balance the pulmonary and systemic circulations, thus facilitating the nursing management. Diastolic blood pressure is also better maintained than in the case of a conventional shunt. In our experience, these patients tend to be haemodynamically more stable in the postoperative period, and tend to tolerate adverse events like excessive bleeding better than do their conventional counterparts. Though our experience with this technique is as yet limited, we believe it is a promising technique that may go a long way in simplifying postoperative nursing management.

DELAYED STERNAL CLOSURE

Leaving the sternum open in the postoperative period can help in the prevention of low cardiac output by reducing the restrictive effects of oedema in the chest wall. It can also provide emergency access should the patient require reexploration of the chest cavity. A baby with an open chest creates special challenges for the nursing team. These babies need to be handled and managed in a special way. Hand washing to prevent infection is the most important factor in nursing the infant with an open chest. Regular swabs must be taken, and prophylactic antibiotics are needed. In those cases where the chest needs to remain open for prolonged periods of time, prophylactic antibiotics can be discontinued and the patient can be maintained on orally administered drugs to decontaminate the gut. Very close surveillance for detecting infections at their earliest, nonetheless, is of paramount importance. The infant is nursed in the supine position, with little scope for relieving pressure. The head may be turned from side to side. Use of gel pads, ripple cooling, and warming mattresses may be of benefit. Pressure-relieving mattresses must be hard enough to provide support to the chest in the event that cardiac massage is needed.

Safety is paramount, and must include having the bed at an acceptable height for the surgeon in the event of emergency intervention. Height-adjustable cots and cot elevators are available. There must be adequate light in the event of emergency chest-opening. The emergency chest-opening trolley should be correctly stocked and nearby at all times. Operating theatre staff should be aware of a child in intensive care with an open chest, and cross-matched blood should be easily available. The defibrillator must have a facility to connect to internal paddles, and a pacing box with appropriate leads should be accessible. Infants can lose body heat extremely quickly through an open chest, and can easily develop hypothermia. The choice of warmer/heater is important, as radiant heat can increase insensible losses and dry out the chest dressings and skin edges. It is important to try and keep the temperature constant. If the infant is allowed to become cold, the peripheral vascular resistance will increase, in turn increasing the flow of blood into the lungs and causing hypotension and pulmonary oedema. Care must be taken to ensure that the infant is adequately sedated and pain free, although some movement may be allowed for the mobilisation of oedema. The chest dressings must be clean and dry, with labels that notify the staff that the chest is open. Dressings should be changed using an aseptic technique. The dressings may be weighed to estimate losses. A Silastic patch is sutured between the skin edges, and it should be observed for signs of tension, bulging, or drying out. If the chest is stented open, this should be recorded because of the implications for cardiac massage. Physiotherapists and radiographers should be informed that the chest is open. Physiotherapy at this stage usually consists only of hand ventilation and suction, especially as the child requires minimal handling to remain stable. Lifting the child for radiography should be done with as little movement as possible, keeping the child level and protecting any lines or pacing wires.

The chest is closed when the infant is stable, has only a small amount of oedema in the chest wall, and requires minimal inotropic support. The surgeon will inform the nurse and the family when the chest is to be closed. It is a procedure that usually takes place in the intensive care unit to minimise the risks of moving the infant back into the operating room. It is the responsibility of the nurse to prepare the child and surroundings prior to the procedure.

The infant needs to be positioned to the right side of the cot with a small roll under the shoulders to extend the neck. It is usual to place incontinence pads under the child to collect any fluids spilt while cleaning the chest, or from bleeding during the procedure. An extension to the central line, in which drugs can be given, needs to be added and brought up to the head of the patient for the use of the anaesthetist. Pacing wires should be connected to the pacing box, and also placed close to the head. Emergency and anaesthetic drugs should be placed near the head of the cot. Blood should be in the refrigerator nearest to the intensive care unit, and other colloids should be drawn up and placed at the head end of the cot. The infant should be clean and dry to allow for good contact of the diathermy pad. Diathermy-compatible electrocardiographic leads should be used. A defibrillator with appropriately sized internal paddles should be placed at the left of the cot. The cot should have all unnecessary objects, such as toys, removed. All furniture should be removed from the cubicle to allow room for the equipment. All unnecessary staff and other visitors should be asked to leave the unit, and the doors kept closed. Nurses who are directly involved and who will need to enter the bed space should be wearing a hair covering and face mask. The infant is now ready for the surgeon to close the chest.

MANAGEMENT OF HAEMORRHAGE AND COAGULOPATHIES

All neonates are at risk preoperatively of developing haemorrhagic disease of the newborn due to deficiency of the clotting factors II, VII, IX and X. If a neonate is admitted to the intensive care unit, the nurse should include in the nursing notes when and if vitamin K was administered. The parents are usually able to give this information because consent is obtained prior to its administration. Prior to surgery, the infant should have samples of blood taken to assess for the presence of a coagulopathy. Prothrombin time, activated partial thrombin time, and reptilase time should all be monitored. A full blood count, which assesses haemoglobin and platelets, should also be obtained. Maintaining a high packed cell volume of between 0.45 and 0.5 and a haemoglobin of 15 g aids transportation of oxygen in the cyanotic baby. In the newborn infant, a sample of maternal blood helps with cross-matching and antibody screening if the infant should require transfusion.

If the infant with duct-dependent congenital cardiac disease should collapse due to closure of the duct, then the circulatory failure that ensues may lead to a complex coagulopathy called disseminated intravascular coagulation. The nurse should be vigilant to the signs and symptoms associated with the disturbance of the clotting mechanisms. The child should be observed for signs of obvious bleeding. This may be noticeable from sites of puncture or placement of venous or arterial lines, in the secretions suctioned from the endotracheal tube or from the gut, in faeces, or in nasogastric aspirate. A clotting profile should be sent to the lab daily, and the results recorded. This may be repeated more frequently if the clinical condition deteriorates. Other less obvious signs of coagulopathy may include purpuric rash, which may be seen underneath blood pressure cuffs. Neurological observation should also be carried out, as neonates are at particular risk of intraventricular haemorrhage.

Management of the diffuse intravascular coagulopathy includes treatment of the basic cause. Administration of inotropes to treat low cardiac output and of antibiotics to treat sepsis is an important part of management. Fresh frozen plasma may be given to replace the consumed clotting factors. It is important to remember that this will need to be given within an hour of defrosting before the clotting factors become inactive. Cryoprecipitate may also be given to increase the levels of fibrinogen. As with all intravenous additives, blood products should be carefully checked against the case notes and identity bands prior to administration, and the cross-match form should contain the pack details, including the number of the issued unit. The patient should be observed for signs of adverse reactions, including rash and pyrexia. If levels of vitamin K continue to be low, then intravenous vitamin K should be given to replenish depleted stores.

In the postoperative period, the most important consideration for the nurse is to manage bleeding. The infant will return from the operating room with one or more chest drains, which will need to be unclamped, placed below chest level, usually into drain stands on the floor, and measured immediately. The infant is then placed on low-grade suction, at around 5 kilopascals, and can also be gently milked to maintain patency. The drainage needs to be measured at frequent intervals, even as often as every 10 minutes in the infant who is draining more than 5 mL/kg/h. The type of drainage should be noted. Thick, red, and warm drainage indicates fresh bleeding. Losses through the drains need to be replaced to prevent hypovolaemia, which in turn could lead to shock, acidosis, and cardiac arrest. Drainage is usually replaced in response to the child's cardiovascular parameters, and is based on measurement of central venous pressure, blood pressure, common atrial pressure, and the core peripheral temperature gap. It should be infused slowly to prevent the peaks and troughs in filling pressures, which can lead to instability.

The choice of colloid replacement is dependent on the packed cell volume. If the packed cell volume is less than 0.34, then blood is used to replace any drainage. If the packed cell volume is greater than 0.36, then plasma is used. If bleeding is persistent, then fresh frozen plasma is transfused, and protamine may be considered. Once the clotting is within normal limits, manufactured plasma may be given. The clotting and platelet count should be checked regularly. Platelets may be low in the postoperative period as they are often caught up or macerated in the filters of the bypass circuit. Cardiopulmonary bypass also interferes with platelet function, so bleeding can occur even in the presence of adequate platelet counts. Platelets should be readily available for transfusion and should be given using the appropriate set. Blood loss into the dressings should also be considered, and can be estimated by weighing the chest dressings.

Cardiac tamponade is a potential postoperative complication following surgery. Most patients have deranged clotting following cardiac surgery, and have some loss of blood. It is important that the nurse should observe closely the amount of chest drainage. Drains that have been draining a modest amount and that suddenly become dry should be treated with caution and their patency checked. The signs and symptoms of cardiac tamponade include cooling limbs, decreasing or absent urinary output, tachycardia, low blood pressure, and a rising atrial pressure. In the child with an open chest, the Silastic patch may be inspected for signs of tension or bulging. These signs and symptoms may occur slowly over several hours, or very quickly over minutes. It is important that an emergency chest-opening trolley be kept in the ward area and checked daily.

All infants who have undergone the first stage of the Norwood protocol have a systemic-to-pulmonary arterial shunt. This requires anticoagulation therapy to prevent it from being thrombosed. Initially, intravenous heparin at 10 U/kg/h is given once bleeding following the operation has settled. When the chest has been closed and milk feeds have been established, it is safe to commence treatment with aspirin.

THE MANAGEMENT OF FLUID BALANCE AND THE KIDNEYS

Maintaining an accurate fluid balance is essential in both the pre- and postoperative periods. The overall aim is to achieve homeostasis, thus preventing fluid overload or hypovolaemia and their associated complications. The nurse needs to assess the patient for signs of fluid overload. The infant may have generalised oedema, seen periorbitally, or in the chest wall, scrotum, or limbs, and a bulging fontanelle with separation of the sutures. There may be signs of systemic venous congestion. The liver may be palpable, with obvious ascities. Endotracheal secretions may become loose, copious, and pink, indicating pulmonary oedema. The skin may become shiny and taut, promoting breakdown of the tissues and pressure sores. Central venous pressures may be elevated. Blood tests may indicate hypoalbuminaemia. Signs of hypovolaemia are generally manifested as signs of low cardiac output as detailed earlier in the chapter. Regular assessment of urea, electrolytes, and lactate must be undertaken in these infants.

Prerenal failure is the most common type of complication seen in infants with congenital cardiac disease. This is caused by low cardiac output and periods of hypoxia and hypovolaemia. If the urinary output should diminish, electrolytes will not be adequately excreted, which results in high levels of urea and creatinine, and hyperkalaemia. If diuretic therapy fails to improve the situation, supportive renal therapy is required. This is usually achieved using peritoneal dialysis. It is standard practice in some units to place a dialysis catheter in the peritoneal cavity after a Norwood procedure.

NEUROLOGICAL MANAGEMENT

Preoperative neurological assessment is vital to establish the presence of any underlying abnormality. Around one-third of these infants have a congenital problem involving the central nervous system.[3] Identification of further problems for these infants may affect the decisions made by the family and medical team with regard to future treatment. Establishing a preoperative baseline assists in the assessment of neurological function subsequent to bypass. The infant's pupils should be examined for equal reaction to light, and this should be recorded in the nursing documentation. Tone, limb movements, and general posture should all be observed. The medical staff should also check reflexes according to age and stage of development, palpate the anterior fontanelle for signs of tension or bulging, and organise ultrasonic cranial evaluation. Observation of the child for siezures is also important.

A recognised complication of cardiopulmonary bypass is the risk of embolisation of air or particulate matter. It is important that the infant be assessed regularly in the postoperative days for neurological injury, especially if there has been any period of prolonged hypoxia or hypoperfusion, or a period of circulatory arrest during the Norwood procedure. Assessment may prove difficult due to the administration of sedatives and paralysing agents. If the baby has to be systemically cooled to manage postoperative junctional ectopic tachycardia, then it must be remembered that hypothermia reduces the level of consciousness.

Neurological assessment also assists the nurse in ensuring that the infant is appropriately sedated and is receiving adequate analgesia. Constant assessment, management, and evaluation of the baby for pain is paramount. Ideally, a pain-scoring tool is used that incorporates both behavioural and physiological cues.[12,13] Signs of pain in infants and children can be assessed by observing facial expression, crying, body movement,[5,14] heart rate, sweating, blood pressure, and saturations of oxygen.[15,16] If the baby is ventilated and receiving sedation, then assessment becomes much more difficult, especially if a paralysing agent is in use.

The management of pain is usually pharmacological. Intravenous fentanyl is the drug of choice due to its analgesic and sedative properties. Fentanyl also helps to dull the response of the pulmonary vascular bed, and therefore helps to prevent changes in pulmonary vascular resistance in response to noxious stimuli.[5] Non-pharmacological interventions include diversional therapy such as music, stories, mobiles, and visual toys, and the use of touch as in stroking and petting. It is very important that the parents be encouraged to talk to their baby.

MANAGEMENT OF THE GUT

Patients who have a life-threatening illness requiring intensive care have increased metabolic demands, which may greatly increase their chances of becoming malnourished. Enteral feeding should be attempted as early as possible because it has been shown to reduce the response to stress postoperatively, to encourage the resumption of gut motility, and to reduce the risk of translocation of bacteria from the gut.[16] If the mother is expressing breast milk, this should be given. All infants who are sick enough to require intensive care are at risk of developing pressure trauma. This risk is reduced in those infants who are well nourished, with normal saturations and good tissue perfusion. As these infants are already coping with low saturations of oxygen and are at risk of poor perfusion, it is important to concentrate on ensuring that they have an appropriate intake of calories and nutrients. The best way to provide this is to try to establish enteral feeding as soon as possible. Due to poor perfusion of the gut and the risk of necrotising enterocolitis, this is not always easy to achieve. The presence of umbilical lines precluding feeding is under debate. For infants who are not able to feed enterally, total parental nutrition is considered at an early stage in their care. The nurse should make a regular assessment of the function of the gut by monitoring gastric aspirate from the nasogastric tube and the bowel actions, both of which should be checked for the presence of blood.

Feeding should be initiated as soon as is possible. If the baby is not absorbing feeds, then the abdomen should be assessed for signs of increasing abdominal girth and auscultated for the presence of bowel sounds. An abdominal x-ray may be ordered for assessing the possibility of necrotising enterocolitis. If necrotising enterocolitis is suspected, the infant needs to be observed closely for signs of severe infection and for increasing metabolic acidosis and rising lactate.

SEPSIS

Infection is an important complication to be avoided in both the pre- and postoperative phases of care. All staff, parents, and visitors must implement good standards of hygiene and hand washing.[17] Outside garments should be removed before coming into contact with the infant. For all infants, both preoperatively and postoperatively, the nurse should take a throat and rectal swab in order to check for bacteria and infection that requires treatment.[18] If it is considered that the bacteria are at risk of translocating out of the gut, the clinician may decide to selectively decontaminate the digestive tract.[19] All infants who have delayed sternal closure are commenced on selective decontamination of the digestive tract. This consists of a combination of tobramycin, colistin, and amphotericin administered orally, with amphotericin also used as a paste applied around the mouth. Prophylactic intravenous antibiotics, which protect against both gram-negative and gram-positive bacteria, are administered at induction of anaesthetic and are continued for 5 days unless the infant is extubated. The infant is then carefully observed for signs of infection. Blood cultures are taken if clinically indicated.

GENERAL CARE

Infants with hypoplasia of the left heart also require the generalised care required by all babies, including assessment of their eyes, mouth, diaper area, cord, and pressure areas. It is important that these aspects of care are implemented as appropriate to the condition of the infant. A child with severe congenital cardiac disease will not tolerate constant handling, and can deteriorate in condition very quickly with something as simple as a diaper change. Raising the legs above the head to change the diaper can cause extra blood to return to the lungs, and make the baby quite unstable. It is important that this is explained to the parents and family in a sensitive way and to include them whenever possible in the basic care of their new baby. This may mean negotiating times for care to take place that are appropriate for the infant, the family, and the multidisciplinary team.

In the infant with hypoplasia of the left heart, any activity that causes the skin to cool has a direct influence on systemic vascular resistance. It is important to try and maintain a constant ambient temperature, even when exposing the infant for examinations or a diaper change. Unnecessary bed-bathing and changing sheets can put the infant at risk of deterioration. It is important to try to complete all activities with the minimal of fuss and disturbance, and with awareness of the environmental temperature. The use of heat shields to exclude drafts is appropriate. In the very unstable infant, it should be remembered that solutions, such as baby lotion, that are stored at room temperature might feel cold. Following any nursing care or medical examination, we should expect that the infant would need time to return to a stable condition.

CONCLUSION

Caring for babies with hypoplasia of the left heart is a nursing challenge. Babies are extremely delicate after the first stage of the Norwood procedure. Subtle changes in parameters may be indicative of serious underlying events. Prompt detection of these subtle changes results in timely intervention, and prevents potentially disastrous complications. As the initial responder to such events, the nurse is one of the most important members of the multidisciplinary team caring for these patients.

References

1. Sano S, Ishino K, Kawada M, et al. J Thoracic cardiovascular surgery 2003 Aug;126:504–9.
2. Charpie JR, Kulick TJ. Pre and postoperative management of patients with hypoplastic left heart syndrome. Prog Paediatr Cardiol 1996;5:49–56.
3. Glauser TA, Rorke LB, Weinberg PM, et al. Congenital brain anomalies associated with hypoplastic left heart syndrome. Paediatrics 1990;85:984–990.
4. Jobes DR, Nicolson SC, Steven JM, et al. Carbon dioxide prevents pulmonary overcirculation in hypoplastic left heart syndrome. Ann Thorac Surg 1992;54:150–151.

5. Hansen DD, Hickey PR. Anaesthesia for hypoplastic left heart syndrome. Anaesth Analgesia 1986;65:127–132.
6. Norwood WI, Lange P, Hansen DD. Physiologic repair of aortic atresia–hypoplastic left heart syndrome. N Engl J Med 1983;308:23–26.
7. Boothroyd AE, Murthy BVS, Darbyshire A, et al. Disseminated intravascular coagulation-diagnosis and treatement. BMJ 1997;312:683–687.
8. Borsattino-Callow L. Current strategies in the nursing management of infants with hypoplastic left heart syndrome undergoing first palliation with Norwood operation. Heart Lung 1992;21(5):463–470.
9. Livermore A, Kotecha S. Nitric oxide for the newborn infant—past, present and future. Care Crit Ill 1999; 15(5):157–162.
10. Rykerson S, Thompson J, Wessel D. Inhalation of nitric oxide: an innovative therapy for treatment of increased pulmonary vascular resistance. Nurs Clin North Am 1995;30(2):381–390.
11. Greenough A. Nitric oxide—clinical aspects. Care Crit Ill 1995;11(4):143–146.
12. Horgan M, Chanoora I. Measuring pain in the neonate: an objective score. Paediatr Nurs 1996;8(10):24–27.
13. Krechel SW, Bildner J. Cries: a new neonatal postoperative pain measurement score—tests of validity and reliability. Paediatr Anaesth 1995;5:53–61.
14. Franck LS. National survey of the assessment and treatment of pain and agitation in the neonatal intensive care unit. J Obstet Gynaecol Neonat Nurs 1986;16(6): 387–388.
15. Owens ME and Todt EH. Pain in infancy: neonatal response to heal lance pain. Pain 1984;20:77–86.
16. Shreeran M. Pain in infants: a literature review. J Neonat Nurs 1997;Nov(Volume 3):13–18.
17. Quaraishi AZ, Mcguick M, Blais FX. Duration of hand washing in intensive care units. A descriptive study. Am J Infect Control 1984;12:83–87.
18. Van Saene HKF, Stoutenbeck CC, Satoller JK. Selective decontamination of the digestive tract in the intensive care unit. Current status and future prospects. Crit Care Med 1992;20:691–703.
19. Sun X, Wagner DP, Knaus WA. Does selective decontamination of the digestive tract reduce mortality for severely ill patients? Crit Care Med 1996;24(5): 753–755.

Chapter 12

The Psychosocial Problems Faced by the Families of Children with Hypoplasia of the Left Heart

Suzie Hutchinson

While considering the diagnosis and treatment of hypoplasia of the left heart, it is important also to look at the psychological, developmental, and social problems that the children and their families face. As previous chapters have discussed diagnosis and treatment, this chapter discusses the needs of the family, especially the parents, during the initial diagnosis and treatment of the condition, explores the problems confronting the children as they develop, and examines some of the issues that affect those with special needs living within the community.

INITIAL DIAGNOSIS

To most parents, the realisation that their child has hypoplasia of the left heart, be it diagnosed antenatally or postnatally, is a devastating shock. The expectant family is anticipating a normal child. The pregnancy, until diagnosis, will have been normal, with the baby growing as expected, and the mother feeling healthy. All the early tests done during pregnancy will have been negative, reassuring the family that all is well.

Nowadays, of the 200 babies diagnosed in the United Kingdom each year with hypoplasia of the left heart, approximately half are diagnosed at the ultrasonic scan performed at their obstetric centre between 18 to 20 weeks' gestation, and then further confirmed at their regional fetal cardiac centre.[1] At this stage, the most difficult problems for parents and families are trying to understand the diagnosis and the options for treatment, whilst at the same time trying to cope with the emotional trauma that goes with their feelings of loss. Many families exhibit the same emotions as families who are bereft. They often become angry and frustrated, denying the diagnosis and demanding a second opinion. Then they show signs of depression and weepiness. One mother described the experience as "being in a pot of treacle—the more you struggled to the edge, the harder the treacle pulls you down."

Amongst all this turmoil, the expectant couple are expected to make a decision about whether to continue with the pregnancy or interrupt it. Many families ask to be led through this mire of feeling. Although it is extremely important to support and inform them, families must make the decisions themselves. During this time, therefore, it is important that families are offered the chance to talk through the diagnosis and options for treatment more than once, and with different clinicians involved in their care, for example, cardiologists, specialist nurses and midwives, fetal medical obstetricians, and possibly surgeons. Families welcome the use of visual aids during these consultations, ideally ones that they can keep for future use. They should always be given the opportunity to return with the questions that such discussions generate. To help complete their personal information jigsaw, they also need to be offered the opportunity to talk to other couples and families that have had to make the same choices. Two charities based in the United Kingdom, namely Little Hearts Matter (www.lhm.org.uk), a support and information service for families of children in whom a single functioning cardiac ventricle disorder has been diagnosed, and Arc (www.arc-uk.org) who offer support to families that have received antenatal results and have been given choices for treatment. These and other organisations offer information geared to nonmedical readers and nonjudgemental support from diagnosis onward.

Having made their choices with regard to treatment, couples must be supported, whatever decision has been made. Those who have opted for pregnancy termination must be allowed the opportunity to grieve for their baby, with nonjudgemental support provided throughout. Those who decide to continue with the pregnancy need support and more information about the possible options for the treatments available at birth. Many of these families, although stressed and frightened, have the opportunity to learn and understand more about what is to follow. They build mechanisms for support, drawing on the many professionals and families that they have had the opportunity to meet. These parents have time to collect the information necessary to make an informed decision about consent for surgery. After the birth of one such baby, his mother said to me, "I can now love my baby and not waste any of our time." Little Alex, unfortunately, died. But because his family had the opportunity to plan, they have many mementos of his short life. He is part of their family.

DIAGNOSIS AFTER BIRTH

The opportunities to understand the disorder and to start to prepare for the treatment to come are not available to those families for whom the diagnosis has not been made before birth. For these families, the trauma is magnified, since they swing from the joy of what seems to be a normal delivery and a healthy baby to the devastating discovery that the baby has a noncorrectable condition for which only very complex and risky palliative surgery is available, with the great chance that, if their baby lives through the first 5 years, he or she will probably face heart transplantation as a teenager.

These families may well have had to travel many hundreds of miles for this information. The mother will be sore and tired following a delivery. The father will be trying to keep things together, whilst possibly juggling other children, work, and his own emotional turmoil. They will need to make a decision as to whether to opt for the palliative surgical options or not. They have none of the advantages of time that are now offered the families in whom the diagnosis is made antenatally. They will be feeling even more bereft, confused, and very frightened. One mother described this experience as being "in a fast moving river. Rocks and branches jut out into the water." These "rocks and branches" were the ever-increasing difficulties for their child that they were learning about as the clinicians explained the condition, treatments, and life expectancy. The mother said she could hear the waterfall at the end of the river. Unable to free herself from the water, she was hurtling toward the drop. Suddenly, she was flung into the air, falling a vast distance into a pool of deep cold water, which engulfed her. Eventually, arms stretched towards her, and she was able to pull herself free, never quite losing the bone-deep coldness that the water had caused.

This metaphor provides us with an insight into the feelings of the parents, their lack of control, and above all, the impression of inevitability. It is important, therefore, that these families, like the antenatal group, are offered many sources of support and information. The cardiac liaison nurses are able to bridge the gap between the medics and the parents, offering informed explanations and a caring word. The families also need the opportunity to speak to other families who have had similar experiences.

Some families decide that treatment is not for them. They may wish to take the baby home, and be supported until the baby dies, so-called comfort care. Others prefer the support of the hospital, but may like to travel back to their original centre of care. Families need the opportunity to spend as much time with the baby as they can. Other children should be encouraged to visit, and grandparents and friends should be allowed to be part of the baby's short life. The religious needs of the family should be considered, with every opportunity for the family's own support network to visit and be part of care. During this short time with the baby, the family needs to build up memories that they can keep with them forever.

Those families that have opted for treatment now ride the roller-coaster of the surgical progression for palliation of hypoplasia of the left heart, with all the traumas described in Chapter 7. Families now begin to use the resources that they have previously been offered; they look for more information and gradually look forward to what may lie ahead. It is important that they are offered consistent support, building their relationship with the medical and nursing team, and using parent-to-parent support as their way to understanding the life of a child with this complex condition. Often the professional teams have to help a family sift through the information they collect. Access to the Internet can cause as many problems as it solves, with families often subjecting themselves to sensationalised research and journalism that uses language that is difficult to understand and interpret. Directing parents to recognised Web sites and documentation ensures that they have access to information written for them. Later, when there is a fuller understanding of the condition, they will be able to explore the Net without leaving themselves open to misunderstanding.[2–4] As the baby's condition improves, parents need to be introduced gradually into care. This helps them to stay involved, reducing their worries of detachment.

"George wasn't mine whilst he was on intensive care." As this mother's comment indicates, there are great difficulties in parent bonding when a child is seriously ill. Encouraging a mother to express breast milk, which can be used in supplemental feeding, helping a father learn how to do eye care, and suggesting the keeping of written and pictorial diaries can all help families to feel included. The bedside and liaison nurses do a great deal to observe and solve problems of parent–child interaction, working as a team to find comfortable ways to work through any problems. Many families express the concern that, if they get too close to the child and he or she subsequently dies, it will be harder for them to cope. Gently, they need to be encouraged that the reverse applies, that their feelings of loss will be harder to cope with if they fail to establish any relationship with their child.[5]

For some families, especially those from abroad, it is impossible for the parents to travel with their child for treatment. For these parents, there is a great need to record their child's activity. Using pictures, poems, and text, translated if necessary, diaries can be kept so parents have a little piece of the time they were away from their child. Saving identification bands, bootees, and even ventilator hats can help parents feel that they have a memory that otherwise would be lost to them.

During treatment for this condition, there are risks that the child might die at anytime. Families need to be offered the opportunity to express the fears that this knowledge generates. Once parents have talked through these worries, they often make a positive decision that they will lead as normal a life as possible, facing any problems as they arise rather than worrying constantly. Other families become stressed and insular. All the family activity revolves around the sick child, and any normal childhood differences are blown out of proportion, making it difficult for families to remain united. There is evidence that, in some communities, more than two-thirds of families with a chronically disabled child undergo major marital stress leading to divorce.[6]

As parents face the child's discharge, it is important that they take more and more responsibility for the day-to-day needs of their child. It is only then that any worries about their confidence or ability will be observed. Gradually, they will learn to feed, bathe, dress, and medicate their child. This needs to be supported by the experienced ward staff, with time being given for families to express their fears and worries, especially if there are still concerns about weight gain, cardiac function, or oxygenation.

Hand-held records or information folders should be used to back up the verbal instructions that are being given. All information should be written in an easily understandable form to ensure that all members of a family can use it. Empowering families with knowledge concerning their child with hypoplasia of the left heart will help to ensure better observation and reaction should the child become sick. This helps families approach day-to-day care with confidence, whilst also providing a better, more logical, response when needed.[7]

LIFE AT HOME

Discharge from the hospital to home causes many stresses and strains on the family of a child with hypoplasia of the left heart. Adjusting family life to any new arrival is difficult, but when the child has a life-threatening disorder the adjustments may take a great deal of time, and for some it will never quite happen. Finding themselves miles from their centre for tertiary treatment, within a community where the medical team knows little or nothing about the condition of their child, increases significantly the level of stress placed on the family, making it even harder to cope at home.

Finding the balance between sensible concern about their child and overpowering worry is very difficult for many parents. Often, the period between the first and second stages of palliative surgery, that balance is never reached. After the second operation, if things have gone well, a more settled daily routine emerges, interjected with only brief periods of hospitalisation and treatment. There are many areas of life at home, nonetheless, that remain to be explored. Below, I highlight some of the most common parental worries, offering some ideas as to how the problems might be alleviated.[8]

THE COMMUNITY TEAM

As previously mentioned, one of the greatest fears expressed by parents is that their family practitioner and visiting nurse do not understand their child's condition. In this respect, it is important to remember that it would be unreasonable to expect every community doctor and nurse to understand the extremely complex and relatively rare condition of hypoplasia of the left heart. They are trained to look after healthy children who may occasionally become ill, rather than children who have complex congenital cardiac disease requiring long-term medical support.

To help the transition from hospital to home, the hospital liaison and the nurses based on the ward of the tertiary centre should make contact with, and work with, the community team, ensuring a basic understanding of the condition and providing a route through which they can express concerns. The use of hand-held records or information sheets helps to ensure continuity of care. On the child's discharge from the hospital, families are advised to make a routine appointment with both their family practitioner/local paediatrician and their visiting/community nurse, to introduce these professionals to the new baby, and to permit them to take a baseline set of observations, for example, weight, height, colour, respiratory rate, and a list of current medications. It also gives both the parents and the medical staff the opportunity to express any concerns that they may have about community care. They can then build a structure for ongoing support and access to swift medical help should it be required. There are many areas of care that parents have to learn to take on themselves following the child's discharge from the hospital.

Heart Failure

Having been surrounded by clinical support within the hospital, many parents worry that they will not know if their child's condition has deteriorated once they have been discharged from the ward. Understanding the signs to look for, and knowing what to do about them, can help families cope with the feelings of medical isolation at home. Before discharge, the signs of heart failure should have been explained to parents with the use of hand-held records and information sheets. There are various publications now available to help parents understand the diagnosis and treatment of these complex conditions, and details are available on the Web site of Little Hearts Matter (www.lhm.org.uk). It is often reassuring to provide contact numbers that parents can call if they have any worries. There may be many calls in the first few weeks, but gradually, as the child settles and parents become more confident, the need to call decreases. The family practitioner and the visiting nurse should be used as contact points, since any ailments might be simple and easily treated by them.

Diet and Weight Gain

For many parents, and their community team, diet is the area of care at home that causes the most consternation. Children with hypoplasia of the left heart have a very high nutritional requirement. Their bodies run inefficiently, so they need many calories just to maintain the system. To ensure growth, they need added calories. However, they tire easily when feeding, and they also need careful fluid balance to ensure that they do not tip into uncontrolled heart failure. In this respect, parents need a great deal of support during their first few months at home, with lots of encouragement with feeding even if their baby gains only small amounts of weight. There is no reason to dissuade mothers from breast-feeding if they and the baby are happy with it. Extra calories can be given by mouth just before feeding, or supplementary feeds can be provided if the baby is not gaining weight.

Mixed feeding often ensures satisfactory gains in weight over the first few months. Oral feeding can be encouraged in the day, with overnight continuous nasogastric or gastric feeding helping to raise calorie intake at a time of least activity. Although charts for weight are used for children with hypoplasia of the left heart, they are only one way of assessing the gain, and should not be used to compare weights with the "normal" scale for that age group. Often parents feel so responsible for their baby's failure to gain weight that they lose the joy of having their child at home. Meals become a battle, and families become extremely tired and tense. This adds to the strain, making the circle of feeding a vicious one. It is important that the hospital and community teams work together with families, listening to the needs of parents, grandparents, and siblings, encouraging and praising even the smallest weight gains. Realistic targets can be set to ensure that plans for feeding fit into family life.

After the second stage of palliative surgery, feeding often settles down. Children wean at this stage, and so it is easier to give them more calories. They will gradually have more energy and become less breathless because of the surgical rerouting of blood to the lungs. These children are very unlikely ever to be overweight, but gradually they will settle into a feeding pattern of a small amounts of food at regular intervals, eating higher calorie foods, and enjoying eating with their family.

Development

It is very difficult in the early days of childhood to make judgements as to how a child will develop. When those days may have been interrupted by hospitalisation, bypass surgery, and periods of time under artificial ventilation, development becomes even more difficult to assess. The majority of children, nonetheless, appear to have developed normally. They are able to walk, talk, ride bicycles, and climb ladders. Occasionally, there are problems with either gross or fine motor skills, or social developmental delay. The evidence at present is that these cases are few and far between (see Chapter 13).

Some children experience increasing fears on hospitalisation. These often manifest themselves as phobias against needles or doctors. Much can be done to alleviate these fears. One option is to ensure that parents have a good understanding of treatments, so that they have as much control through understanding as they can, thus alleviating some of their stress and consequently reliving some of their child's fears. It also helps to prepare children before a hospital admission through the use of books and games, coupled with a preadmission visit offering the opportunity to discuss their fears, thus greatly reducing the extreme reactions to hospitalisation experienced by some children.

Financial Concerns

For many families, the financial strain of having a child with hypoplasia of the left heart is an added stress. Often there is a need for parents to take extended periods of leave from work, which adds to the extra expenses created by visits and stays in the hospital, as well as the expenses of day-to-day living with a disabled child. To help alleviate this worry, they may be able to apply for various government or state allowances. They often need a great deal of professional help to make a claim, since the application system is complex and off-putting.

Allowances vary by country. Some disability living allowances are available, as are allowances that financially support the caregiver and help with mobility where physical ability is restricted. Individual grants are available to help families that are unable to purchase the medical equipment necessary to keep the child safe or well cared for, or even to acquire a telephone or washing machines or to take driving lessons. It is important that families are encouraged to apply for such allowances, as acknowledgement of a disability can help open doors of opportunity for a child, as well as helping to alleviate the financial strain discussed above.

Education

The thought of their child moving from the safe environment of home into nursery, and then full-time education, fills many parents with dread. Is the child going to be safe? Will people understand his or her needs? Will the child achieve at the same rate as other children? A majority of children with hypoplasia of the left heart cope well with normal mainstream education. They mix well with other children, and develop as children of their age should. Most children need simple support, for example, easy access to the toilet, the opportunity to stay inside if the weather is bad, extra time to get from class to class, and extra care during physical education.

In elementary and junior high schools, the goal is to treat the children as normal individuals. The needs of children with hypoplasia of the left heart are met easily within the school structure. The needs of some children, nonetheless, go beyond the bounds of normal support, and provision of a classroom assistant or specialist educational support is required.

To ensure that individual needs are met, a statement of educational needs is written, including basic physical and educational assessment, as well as observations from the parents and teachers about the ability of the child. The assessment gives the education team the information needed to meet the individual needs of each child. Many families choose to have the assessment done early in the educational life of their child, as it makes it easier to ensure that support continues throughout their education. Schools in the United Kingdom, for example, are unable to fund extra classroom support without the backing of the education authorities, so "statementing" provides for the individual needs of each child.

THE FUTURE

It is difficult to predict the future for these children with any certainty. From the experience gained from children undergoing surgery in the United States, the evidence is that, ultimately, children with hypoplasia of the left heart may need cardiac transplantation, perhaps in their late teens or early twenties. Some children will require this treatment earlier should their condition deteriorate. For many families, the unsure future for their children is very difficult to live with. Many families have successfully gone through the three stages of palliative surgery, have been able to lead a normal life, and have experienced the fun and games of normal childhood. It is difficult for these children and their parents to face the fact that life may not always be so easy, and that there are many obstacles still to face.

Growing through adolescence, experiencing the beginnings of adult life, and feeling invincible, makes the timing of increased heart failure seem unfair. Professionals who have been working with children with congenital cardiac disease for many years have seen these problems before. However, it will be a new hurdle years from now for many of the children with hypoplasia of the left heart currently in our care.

Acknowledgements

This chapter was written with the aid of comments from cardiac liaison nurses, cardiac nurses, cardiologist and surgeons, community health care teams, and most importantly children and their families who live every day with the diagnosis of hypoplasia of the left heart.

References

1. Brackley KJ, Kilby MD, Wright JG, et al. Outcome after prenatal diagnosis of hypoplastic left heart syndrome: a case series. Lancet 2000;356:1134–1147.
2. Kanneh A. The need to communicate. Nurs Standard 1990;5(5):19–20.

3. Hutchinson SE. Getting to the heart of the matter. Nurs Times 1999;10:55–57.
4. Tucker Swanson L. Treatment options for hypoplastic left heart syndrome: a mother's perspective. Crit Care Nurs 1995;15(3):70–79.
5. Klaus MH, Kennell J. Parent and Infant Bonding. St. Louis: Mosby, 1982.
6. Hayes VE, Knox JE. The experience of stress in parents of children hospitalised with long-term disabilities. J Adv Nurs 1984;9:333–341.
7. Coombes R. From parent to expert. Child Health 1995;2:237–240.
8. Macdonald M. Children discharge from hospital—what mothers want to know. Nurs Times 1988;84(16):63.

Further Reading

NAWCH Charter for Children in Hospital, Action for Sick Children. London: NAWCH, 1984.

Court SDM. Fit for the Future: Report of the Committee on Child Health Services, London: HMSO, 1976.

Department of Health. Child Health in the Community: A Guide to Good Practice, London: Department of Health, 1996.

Department of Health. The Welfare of Children and Young People in Hospital, London: Department of Health, 1991.

Rodin J. Will This Hurt? London: Royal College of Nursing.

Sreeram N. Hypoplastic left heart syndrome: current treatment options and outcome. Curr Paediatr 1996;6:156–161.

Chapter 13

Neuropsychological Functioning and Psychosocial Development in Children with Hypoplasia of the Left Heart

Jacqueline C. Blyth

Each year in Britain, about 200 children are born with hypoplasia of the left heart. These children present with a range of symptoms, which may include cyanosis, raised respiratory rate, breathlessness, poor feeding, and cardiac murmurs. Some children are diagnosed antenatally, but most postnatally, usually within the first 4 weeks of life. When a diagnosis is made antenatally, parents are faced with the difficult choice of deciding whether or not to terminate the pregnancy, or be prepared to subject themselves and their baby to the stress and trauma associated with numerous surgical procedures. This chapter provides clinicians and parents with information that will enable them to consider the potential impact of palliative surgery on the intellectual and neurological functioning of their children should they choose to continue with the pregnancy. As with any other life-threatening illness, diagnosis presents families with difficult choices and decisions to make regarding the long-term physical and psychosocial health and well-being of their child.

At our hospital, staged palliation has now achieved early survival in about half of the babies born with prenatally and postnatally diagnosed hypoplasia of the left heart.[1] This improvement in overall survival following palliation has led to an interest in establishing the psychological impact on these children after surgery.

INTELLECTUAL AND NEUROLOGICAL OUTCOME IN HYPOPLASIA OF THE LEFT HEART

A systematic review of the literature between 1966 and 2001 has identified only 32 journal articles referring to outcomes following surgery for congenital cardiac disorders.[2] Of these articles, 28 were related specifically to the outcomes of children with hypoplasia of the left heart, whereas the remainder grouped small numbers of children with this constellation with other congenital cardiac lesions. In only four articles, however, was consideration given specifically to neurodevelopmental outcome of children with hypoplasia of the left heart. Whilst these studies provide some information regarding cognitive functioning following surgery, they also raise a number of methodological concerns. These include inappropriate test measures, little consistency with respect to diagnosis or stage of surgical procedure, and, in some cases, failure to include an appropriate control group. None of these studies assessed the quality of life of children following their surgery.

To assess intellectual functioning, Kern et al.[3] compared the cognitive abilities of 10 children with hypoplasia of the left heart and their siblings. The children were administered the Wechsler Preschool Primary Scale of Intelligence–Revised,[4] the Peabody Picture Vocabulary Test,[5] and the Beery Visual Motor Integration Test.[6] The authors reported no significant difference in results between the children and their sibling controls. In another study, Mahle et al.[7] found that over one-third of children with hypoplasia of the left heart obtained full-scale scores for intelligence in the low range. Mahle et al., however, were less rigid in their criteria for inclusion, and incorporated children with both hypoplasia of the left heart and its variants. In addition, although the patients had all undergone surgery at the same institution, they were assessed at different stages of palliation, some as early as 3 months subsequent to surgery. This has a number of implications. For example, research into children undergoing liver transplantation indicates that there is a noticeable decline in intelligence quotient immediately after surgery, which may continue for up to 6 months.[8] It is neither appropriate nor reasonable, therefore, to make predictions regarding intellectual functioning based on such an early evaluation.

Rogers et al.[9] examined the neurodevelopmental outcome of 11 children with hypoplasia of the left heart using a variety of tools. They included children with severe neurological conditions, including one with cerebral palsy and another with quadriplegia, as well as children at various stages of the three-stage Norwood procedure. According to their findings, two-thirds of children with hypoplasia of the left heart had major developmental disabilities based on descriptive reports, ranging from "suspect" to "severe-profound."

The most recent assessments of neurodevelopmental outcome in children with hypoplasia of the left heart have been conducted by Goldberg et al.[10] and Blyth et al.[11] Analysis in the Goldberg study included comparison of a group of patients with another group having a functionally single ventricle and the standard population. Although the children with hypoplasia of the left heart were reported to have significantly lower scores than the children with a functionally single ventricle, Goldberg et al.[10] found that neither subgroup scored significantly differently compared to the standard population. In the study carried out by Blyth et al.,[11] the results indicated that children with hypoplasia of the left heart did no worse on tests of verbal and performance intellectual functioning than children with other functionally univentricular hearts. The cognitive outcome for children with hypoplasia of the left heart, nonetheless, was significantly worse than their sibling controls. As with other reported studies, such as that of Ehyai et al.,[12] this report also identified a number of risk factors associated with poor neuropsychological outcome. These include evidence of clinical seizures, length of stay in intensive care, and, to a lesser degree, the duration of cardiopulmonary bypass.

Whilst one or two studies have therefore demonstrated a significant reduction in the intellectual functioning of children with hypoplasia of the left heart, there are a variety of methodological inconsistencies in the studies, resulting in a lack of uniformity in methods of assessment. Because of this, it is difficult to compare the outcome of any one study with another. To begin with, the term "hypoplasia of the left heart" is often imprecisely defined, and is frequently used interchangeably with a variety of other terms.[13] Also, many studies that have examined the intellectual or neurological outcome of children following cardiac surgery were based on a comparison of different surgical techniques. These have included extracorporeal membrane oxygenation,[14] duration of circulatory arrest,[15] hypothermic circulatory arrest or low-flow cardiopulmonary bypass,[16–18] different types of intervention,[19] and different malformations.[20–23] Some studies have compared neurological functioning at different stages of development.[24,25] It is also noteworthy that delayed motor development and failure to thrive have been found in children with congenital cardiac malformations,[26] as have perceptual and visual-spatial difficulties, particularly in those children with cyanotic lesions.[27]

Furthermore, apart from demonstrating significant differences in the intellectual and developmental capabilities of children with congenital cardiac disease, these studies of outcomes have also highlighted the dearth of research into the quality of life and adaptability of the children examined. It is important to remember that intellectual development is greatly dependent on a variety of both psychological and psychosocial influences. This raises the issue that as well as measuring the intelligence quotient of children with congenital cardiac disease, it is worthwhile to gain as much information as possible regarding the daily fears and challenges faced by the children. As with other chronic illnesses, children with hypoplasia of the left heart experience many stresses, including altered physical appearance, demanding regimes of treatment, absences from school, and lack of stamina. In addition to these stresses, they must also cope with the typical tasks of childhood, such as developing a sense of autonomy, establishing relationships with their peers, and the normal transitions associated with growing up.[28]

BEHAVIOURAL ADJUSTMENT AND DEVELOPMENTAL DELAY

Apart from the neuropsychological and intellectual concomitants of congenital cardiac disorders, researchers have also investigated the developmental and behavioural characteristics of these children. According to Ferry,[26] developmental problems result from damage to the central nervous system, which occurs with the chronic hypoxia associated with cyanotic cardiac lesions. Samango-Sprouse and Suddaby[29] suggest that children born with congenital cardiac malformations are at risk of developmental delay from birth because the functioning of their organs is reduced immediately by a reduction in the flow of oxygenated blood.

Based on the hypothesis that diminished arterial oxygen saturation may be causally related to slow development, prolonged reaction time, and low intelligence quotient, developmental delay and physical underdevelopment have been documented in children with a range of congenital heart diseases, especially those with cyanosis.[30,31] Other studies have reported an increase in the incidence of emotional and behavioural problems of children with congenital heart disease,[11,32] a higher rate of depression and anxiety compared to healthy peers,[33] and difficulties in social relationships and adjusting to school. There is also some evidence that, during adolescence, alterations in body image may occur, as well as further difficulties in establishing relationships with peers and developing independence.

DIFFERENCES IN GENDER

A number of studies have found differences between the sexes in verbal and performance functioning of children with congenital cardiac defects. Honzik,[34] for example, found lower verbal function and higher scores for intelligence quotient in girls with congenital cardiac malformations. She argued that the nervous system of girls is more sensitive to the stresses associated with congenital cardiac lesions than that of boys, although it is not clear why this should be so. According to Cravioto,[35] boys, but not girls, with congenital cardiac defects have significantly lower intelligence quotients than the normal population. Other reports, in contrast, found no statistically significant differences between the scores for intelligence of boys and girls, regardless of their cardiac condition (e.g., Blyth et al.[11]).

SOCIOECONOMIC AND ENVIRONMENTAL FACTORS

Another factor believed to affect development is socioeconomic status. Children belonging to higher socioeconomic groups have been found to fare better on several developmental parameters, especially language and cognitive skills.[36] Other environmental factors, such as the stresses of early surgery and overprotective family attitudes, may also contribute to developmental lags or deficits.[37]

According to McCall,[38] a substantial proportion of individual differences may be due to environmental experiences that are not shared by members of a family. The suggestion is that as children begin to act on their environment, they produce responses that are matched with their abilities,[39] and as they mature they become more inclined actively to select experiences that match their abilities. This process results in an interplay among factors such as biochemical, biological, physiological, cultural, and societal influences, which play an important role in maintaining their abilities.[40]

Based on this premise, the suggestion is made that as a consequence of their cardiac condition some children with hypoplasia of the left heart may also feel very vulnerable. This, in turn, may produce parental overprotectiveness, leading to poor psychosocial adjustment. In other words, it is possible for psychopathologic changes to occur as a result of the exposure of the children to, and the interaction with, their environment. It is also believed possible for the psychological, social, and emotional problems experienced by some children not to be directly related to their cardiac condition, but more so to their individual experiences and backgrounds. Early studies of children with congenital cardiac malformations have demonstrated an effect on functioning of the families, including disrupted relationships within the family and emotional disturbance. It is believed that infants with cardiac malformations are less likely to be securely attached to their mothers, and they display higher avoidant behaviours and greater dependency than other groups used for comparison.[41] Children with complex cardiac disease have been rated by their parents as more withdrawn, having more social problems, and engaging in fewer activities, and by their teachers as more withdrawn.[42]

According to Kong et al.,[32] problems such as these may be the result of maternal anxiety and various other emotional, social, and psychological problems associated with caring for a child with a life-threatening illness. Majnemer and Limperopoulos[37] suggest that maternal guilt and anxiety together tend to promote behavioural and social problems in children with cardiac malformations. Families of the children have also reported experiencing more stress.[42] Mothers, in particular, have demonstrated an increased sense of guilt, insecurity, and uncertainty about the future, and a sense of helplessness.[43] They may also present with evidence of problems of mental health, such as stress, anxiety, and depression.

NEUROPSYCHOLOGICAL PROBLEMS ASSOCIATED WITH CONGENITAL CARDIAC DEFECTS

The term "neuropsychological" refers to the scientific study of the association among cerebral pathology, mental processing, and behaviour. According to Teeter and Semrud-Clikeman,[44] all behaviour, including cognitive processes, which are essentially psychological, is mediated by the central nervous system and its integrated supporting physiological systems. Mental processes include knowing, thinking, memory, perception, attention, problem solving, humour, learning, motivation, and understanding. Based on this integrated framework of intellectual functioning, the development of the system has a direct impact on the behavioural, cognitive, and psychosocial adjustment of children.

Cognitive development in children is subject to a number of influences, both pre- and postoperatively.

For example, because glial proliferation and myelination in the central nervous system take place most rapidly during the first year of life, any interference with this process, due to a reduction in cardiac output, may have a negative effect on neurodevelopment.[45,46] There is evidence that some children with congenital cardiac malformations have been found to have an increased prevalence of impairment of the central nervous system,[47,48] neurodevelopmental deficits,[16] and structural cerebral lesions and abnormalities.[49]

Following surgery, medical complications such as cardiac arrest and thrombosis may also result in injury to the central nervous system.[50] Neuropathological injury can occur for a number of reasons such as microembolization, air embolism, hypoxia, hypoperfusion, or biochemical and metabolic disturbance.[50–53] Many children with hypoplasia of the left heart become cyanotic at some stage, a feature that has also been linked to developmental delay.[30] Studies of cognitive development indicate that children with cyanotic lesions are at risk for lower intelligence than children with acyanotic malformations, particularly if the disease is severe and if corrective surgery is not performed within the first few years of life.[16]

As with other congenital cardiac disorders, a proportion of neuropsychological dysfunction in those with hypoplasia of the left heart may result from perioperative events, including the support techniques used to support vital organs during surgery. In the past, neurological damage was more likely to occur as a direct consequence of hypothermia, as the result of uneven cerebral cooling, or as a result of a sudden and excessive alteration in temperature between blood and neural tissue. Deep hypothermic circulatory arrest and low-flow cardiopulmonary bypass were also believed to cause marked alteration of cerebral metabolism, resulting in altered cerebral function, and occasionally irreversible cerebral injury.[48] More recently, however, with the growing technological and medical advances in open-heart surgery, hypothermia is now believed to be neuroprotective by reducing metabolic demand, and thereby permitting organs to survive the ischemic insult of circulatory arrest for longer periods than would be possible with normothermia.[36,54]

The events most frequently associated with neuropsychological problems following open-heart surgery are cardiac arrest,[55] deep hypothermic circulatory arrest,[56] seizures,[12] and low-flow cardiopulmonary bypass.[57] The main deleterious effects of cardiac arrest on the brain are caused by a combination of insufficient cerebral blood flow, deficient oxygenation, and anaemia.[58] Other injuries may include damage to organs associated with circulatory arrest, hypoxic-ischaemic damage, encephalopathy, focal vascular lesions, and cerebral haemorrhage, all of which can lead to intellectual impairment such as problems with memory and learning, and speed of processing information.[59] Atypical development of perceptual-motor skills has been found in children with both cyanotic and acyanotic lesions including problems with visual-perception, handwriting, and visual-spatial skills.[29] Surgical intervention is also believed to increase the risk of children developing problems in learning, particularly those associated with distractibility and attention deficit.[60]

NEUROPSYCHOLOGICAL TESTING

Until fairly recently, the majority of tools for neuropsychological assessment for children were based on adult neuropsychology, with little attention paid to the developing brain. More recently, however, the area of developmental neuropsychology has become a distinct subspecialty,[61] with more explicit tests now being developed that provide a breakdown of specific cognitive skills believed to be associated with underlying neurological and cognitive processes. For example, newly developed neuropsychological tests,[62] allow clinical neuropsychologists to carry out multifactorial assessments of neuropsychological functioning, including attention and executive functioning, language skills, sensorimotor functions, visual-spatial processing, and memory and learning abilities.[62] This process-oriented approach also takes into account a variety of qualitative factors, as well as the behaviour of the child during the test.

Neuropsychological testing can provide a standardised and reasonably sensitive method of quantifying the effects of cerebral damage associated with congenital cardiac disease. Whilst some tests may tap only a limited number of neuropsychological functions or systems, the assumption is that they can at least provide some measure of underlying cognitive processes and neurological functioning. Traditionally, neuropsychological testing in children included a variety of psychometric tests, such as the Wechsler age-adjusted scales of intelligence,[63] but recently more explicit tests have been developed that provide a breakdown of different cognitive skills believed to be associated with underlying neurological processes. In terms of making clinical recommendations, however, the diagnostic utility of any neuropsychological test is limited unless additional information is obtained regarding the biological, behavioural, emotional, environmental, and socioeconomic status of the child in conjunction with the possible inherent and treatment effects of the cardiac malformation.

There are a number of inherent difficulties regarding the development of a coherent framework for neuropsychological testing in children, including the constantly changing picture of the developing brain and the interaction between development of the brain and the environment.[64] Other factors also need to be considered, such as what type of damage might have occurred to the brain, which specific area of the central nervous system may be damaged, and what effect this damage might have on cognitive functioning. As emphasised above, in terms of making a clinical diagnosis, the diagnostic utility of any neuropsychological test is limited without much additional information being obtained.

Whilst it is acknowledged that a neuropsychological evaluation provides a careful measurement of the various capabilities of the child to perform, it is worth remembering that, due to the plasticity of the brain, as well as the developmental changes that occur during growth, these tests can measure the response of the brain only at the time the child is tested.

Since the likelihood is that there are some developmental stages at which children are more receptive or vulnerable than at other stages, it would be helpful to demonstrate whether or not there is a critical period during staged surgical reconstruction that would allow clinicians to minimise the impact of surgery. There is also the potential for extreme intraindividual behaviours when assessing children. A child may be emotionally ill-equipped to face what may be perceived as a fearful test situation. Attentional capacity may be quite variable, with the timing of an assessment affecting results, particularly in younger children who may not respond as well if they feel fatigued or tired. Assessing children at different times may reveal different facets of intellectual functioning, and these may differ in the home, at school, and in hospital settings.

To assess the full range of a cognitive abilities, a comprehensive neuropsychological battery is needed that can assess visual and verbal memory, attentional skills, visual-spatial skills, and executive functioning. Other traits and attitudes not measured by standardised tests of intellectual functioning, such as planning and awareness, impulsivity, and persistence, are equally as important in assessing effectiveness and adaptability on a day-to-day basis. Whilst the intelligence quotient may be predictive of performance in the future, Kaufman[65] argues that these scores "should not be accepted as a statement of destiny." It is important to remember that intellectual functioning alone does not predict the ability to engage in future learning and other adaptive behaviours. Therefore, one should be cautious about generalising the results of these tests to other behaviours or performance.

THE FUTURE

At present, the possibility that hypoplasia of the left heart may be associated with neurodevelopmental problems independent of the operative or perioperative is equivocal. Until more conclusive evidence becomes available, it is suggested that the causes contributing to neurodevelopmental, intellectual, and neuropsychological sequels in children with any form of congenital cardiac disease are multifactorial, involving a complex interaction of a range of preoperative, perioperative, and postoperative events. In addition, other environmental factors, such as the stresses of early surgery, long hospitalisations, and overprotective family attitudes, also play a significant part in contributing to the potential for developmental lags or deficits in children with serious congenital cardiac conditions,[49] particularly those with cyanotic disease.

To minimise disability and maximise potential, a number of factors should be considered in the care and management of children with hypoplasia of the left heart. First, it is particularly crucial to ensure that the most appropriate surgical protocols and perioperative medical treatment, as well as assessments for special educational needs, psychological programs, and advice, are made available to the families and teachers of children with congenital cardiac disorders as soon as concerns are raised regarding their functioning.

Second, it is recommended that children with hypoplasia of the left heart undergo early evaluation and careful monitoring throughout their school years. This is of particular importance, since the emerging literature suggests that it is possible to alter some of the adverse effects associated with delayed intellectual development by using highly specific and early programmes of intervention.[44] For example, with early exposure and the teaching of tasks that are affected by the ability to learn, it may be possible to improve performance on those tasks.[51]

Third, because of the wide range of psychosocial factors that impinge on intellectual development, it is recommended that preventive intervention should be considered to help children with hypoplasia of the left heart and their parents to identify the stresses of chronic illness, and help them develop strategies to cope with these stresses in a way that enhances adaptation and daily functioning. Drotar et al.[66] recommend four goals of preventive intervention in child health care: helping children and families cope with the anxiety and fears related to the illness and its management; being aware of the developmental stages in order to facilitate parents' understanding of the illness and improve treatment compliance; assisting families in their attempts to integrate the illness into family life; and supporting families when they need to engage with important systems such as health and education services, as well as social adaptation and integration.

Fourth, as with other serious health-related conditions, many children with hypoplasia of the left heart live fulfilling lives. Assessment of these children needs to move from "What are the difficulties experienced by these children?" to "What are the factors that help children cope with the cardiac malformation?" This means moving away from focusing on the problems these children have to an attempt to try and understand what distinguishes the children who do well from those who do less well. In the future, a study that affords a qualitative view of family members of children with hypoplasia of the left heart, as opposed to only exploring potential cognitive deficits, would contribute to understanding the individual differences of these children in terms of coping, adjustment, and quality of life.[67] Other psychological variables, such as individual personality characteristics, as well as social, family, and school influences, also need to be investigated, since they all interact in mutually influential ways to exacerbate or facilitate intellectual development in children with congenital cardiac disease.[44]

As with other chronic illnesses, children with hypoplasia of the left heart experience many stresses, including altered physical appearance,[68] demanding regimes of treatment, absences from school, and lack of stamina.[69] As mentioned previously, they must also cope with the typical stages of childhood and the normal transitions associated with growing up.[28] As well as focusing on comparisons between physically ill children and healthy children, therefore, it may be just as pertinent to interpret hypoplasia of the left heart as a stress that challenges the resources of the child and family. Other factors associated with parenting, such as a lack of discipline and control,[55] may also account for neurodevelopmental problems, but in the absence of more conclusive research such an inference remains debatable.

Ideally, therefore, a holistic approach is needed to the care of the child and the family, including a psychological assessment investigating mood states, self-esteem, well-being, illness-related stresses, and social support.[70] In this way, it will prove possible to determine the role of interventions in preventing or minimising psychosocial morbidity. This type of assessment requires professionals to have a thorough knowledge of the disease process, the psychosocial aspects of the disease and its effects on children and their families, as well as aspects of child development and the ability of the child and family to cope.

IMPLICATIONS FOR CLINICAL PRACTICE

Based on a policy statement devised by the American Academy of Pediatrics,[71] I suggest that a number of factors should be incorporated in the ongoing assessment and treatment of children with hypoplasia of the left heart. I recommend that a comprehensive cognitive assessment of children should

- be offered as part of a multidisciplinary evaluation;
- be used to inform procedures of treatment and intervention;
- take place at regular intervals throughout the medical and educational career of the child;
- be viewed as part of an overall and ongoing process of assessment and treatment;
- highlight difficulties as soon as possible to facilitate intervention and thereby minimise their impact;
- serve as a reminder to the professionals involved in the care of the child to monitor and observe development;
- be viewed as an efficient way of recording and evaluating information; and
- include a full range of assessments of other cognitive skill, such as visual and verbal memory, attention, executive skills, visual-spatial skills, and language skills.

CONCLUSION

Children with hypoplasia of the left heart are confronted with a variety of problems that are as much psychological and emotional as physical. Their psychosocial adaptation is influenced by many factors, and includes an interplay between their physical, psychological, and social characteristics, as well as their individual stages of development. Whilst psychosocial adaptation is important in its own right, it also has major implications in terms of compliance with treatment and the psychological well-being of the child, as well as other family members.[72] To provide the best possible care, therefore, it is important for clinicians to adopt a holistic approach at all stages of the disease process. The use of a comprehensive clinical perspective, which combines medical, developmental, behavioural, cognitive, psychosocial, biological, and environmental factors, including interaction at school, at home, with peers, and with the family, will provide families and clinicians with the most comprehensive package of care. Because the relationship between these factors is multidirectional, a thorough psychological evaluation of the psychosocial and intellectual functioning of the child allows for the implementation of the most appropriate intervention.

A comprehensive assessment of children with congenital cardiac disease, both pre- and postoperatively, provides clinicians with valuable information regarding evidence of decline in specific areas of intellectual functioning. By identifying potential learning difficulties in these children in the early stages, it may be possible to limit the effect of these problems on their future intellectual functioning and quality of life.

Health care, whether preventative or oriented to treatment, must encompass the psychosocial integrated with the physical, at all phases of the life cycle and at all stages of the disease. If carried out effectively, such care not only improves the general well-being of the healthy members of the family, but also has beneficial effects on the physical treatment and psychological well-being of the child.

References

1. Brackley KJ, Kilby MD, Wright JG, et al. Outcome after prenatal diagnosis of hypoplastic left heart syndrome: a case series. Lancet 2000;356:1143–1147.
2. Blyth J, Jones C. Meta analysis of neuropsychological functioning in children with congenital heart disorders. Unpublished Doctoral Manuscript, School of Clinical Psychology, University of Birmingham, UK: 2001.
3. Kern J, Hinton V, Nereo N, et al. Early developmental outcome after the Norwood procedure for hypoplastic left heart syndrome. Pediatrics 1998;102(5):1148–1152.
4. Wechsler D. Manual for the Wechsler Preschool and Primary Scale of Intelligence Revised (WPPSI-R). San Antonio, TX: Psychological Corporation, 1989.
5. Dunn L, Dunn L. The Peabody Picture Vocabulary Test–Revised. Circle Pines, MN: American Guidance Service, 1981.
6. Beery K. The VMI: Developmental Test of Visual Motor Integration, 3rd Rev. Cleveland: Curriculum Press, 1989.
7. Mahle W, Clancy R, Moss E, et al. Neurodevelopmental outcome and lifestyle assessment in school-aged and adolescent children with hypoplastic left heart syndrome. Pediatrics 2000;105(5):1082–1089.
8. Sexson S, Rubenow J. Transplants in children. In Craven J, Rodin G, eds. Psychiatric Aspects of Organ Transplantation. New York: Oxford University Press, 1992.
9. Rogers B, Msall M, Buck G, et al. Neurodevelopmental outcomes of infants with hypoplastic left heart syndrome. J Pediatr 1994:126(3):496–498.
10. Goldberg CS, Schwartz EM, Brunberg JA, et al. Neurodevelopmental outcome of patients following the Fontan operation: a comparison among children with hypoplastic left heart syndrome, other functional single ventricle lesions and the standard population. J Pediatr 2000:137(5):646–652.
11. Blyth JC, Wassmer E, Ferris S, et al. Neuropsychological and neurodevelopmental outcome of children with hypoplastic left heart syndrome following the three stage Norwood procedure: the British experience. Doctoral research paper presented at the 3rd World Congress of Pediatric Cardiology & Cardiac Surgery, Canada, 2001.
12. Ehyai A, Fenichel G, Bender H. Incidence and prognosis of seizures in infants after cardiac surgery with profound hypothermia and circulatory arrest. JAMA 1984;252:3165–3167.
13. Tchervenkov CI, Jacobs ML, Tahta SA. Congenital heart surgery nomenclature and database project: hypoplastic left heart syndrome. Ann Thorac Surg 2000;69:S170–179.
14. Tindall S, Rothermel R, Delamater A, et al. Neuropsychological abilities of children with cardiac disease treated with extracorporeal membrane oxygenation. Developmental Neuropsychology 1999;16(1): 101–115.
15. Wells FC, Coghill S, Caplan H, Lincoln C, Kirklin J. Duration of circulatory arrest does influence the psychological development of children after cardiac operation in early life. J Cardiovasc Surg 1983;86:823–831.
16. Bellinger DC, Jonas RA, Rappaport LA, et al. Developmental and neurologic status of children after heart surgery with hypothermic circulatory arrest or low flow cardiopulmonary bypass. N Engl J Med 1995;(9):549–550.
17. Clarkson P, MacArthur B, Barrat-Boyes B, et al. Circulation 1980;62.
18. Oates R, Simpson J, Turnbull J, et al. The relationship between intelligence and duration of circulatory arrest with deep hypothermia. J Thoracic and Cardiovascular Surgery 1995;110(3):786–791.
19. Visconti KJ, Bichell D, Jonas R, Newburger J, Bellinger D. Developmental outcome after surgical versus interventional closure of secundum atrial septal defect in children. Circulation 1999;100(suppl II):II-145-ii–150. J Thorac Cardiovasc Surg 1999;110(3):786–791.
20. DeMaso D, Beardslee W, Silbert A, Fyler D. Psychological functioning in children with cyanotic heart defects. Dev Behav Pediatr 1990:11(6):289.
21. Ellerbeck K, Holden W, McMenamin S, et al. Neurodevelopmental outcomes in children surviving d-transposition of the great arteries. Dev Behav Pediatr 1998;19(5):335–341.
22. Mendoza J, Wilkerson S, Reese A. Follow-up of patients who underwent arterial switch repair for transposition of the great arteries. Am J Dis Child 1991;145:40–43.
23. Sharma R, Choudhary S, Mohan M, et al. Neurological evaluation and intelligence testing in the child with operated congenital heart disease. Ann Thorac Surg 2000;70:575–581.
24. Wray J, Sensky T. Controlled study of preschool development after surgery for congenital heart disease. Arch Dis Child 1999;80:511–516.
25. Yang LL, Liu ML, Townes B. Neuropsychological and behavioural status of Chinese children with acyanotic congenital heart disease. Int J Neurosci 1994; 74:109–115.

26. Ferry P. Neurologic sequels of cardiac surgery in children. Am J Dis Child 1987;141:309–312.
27. O'Dougherty M, Wright FS, Garmezy N, Loewenson RB, Torres F. Later competence and adaptation in infants who survive severe heart defects. Child Dev 1983;54:1129–1142.
28. Harbeck-Weber C, Hepps McKee D. Prevention of emotional and behavioural distress in children experiencing hospitalisation and chronic illness. In: Roberts M, ed. Handbook of Pediatric Psychology, 2nd ed. London: Guilford Press, 1995.
29. Samango-Sprouse C, Suddaby EC. Developmental concerns in children with congenital heart disease. Curr Opinion Cardiol 1997;12:19–98.
30. Aisenberg RB, Rosenthal A, Nadas AS, Wolff PH. Developmental delay in infants with congenital heart disease. Pediatr Cardiol 1982;3:133–137.
31. Rosenthal A, Castaneda AR. Growth and development after cardiovascular surgery in infants and children. Prog Cardiovasc Dis 1975;28:27.
32. Kong SG, Tay JSH, Yip WCL, Chay SO. Emotional and social effects of congenital heart disease in Singapore. Aust Paediatr J 1986;22:101–106.
33. Utens EM, Verhulst FC, Meijboom FJ, et al. Behavioural and emotional problems in children and adolescents with congenital heart disease. Psychol Med 1993;23(2): 415–24.
34. Honzik MP. Response to Cravioto J. Sex differences in IQ pattern of children with congenital heart defects. Science 1970;174:1043–1044.
35. Cravioto J. Sex differences in IQ pattern of children with congenital heart defects. Science 1970;174:1042–1043.
36. Blackwood MJA, Haka-Ikse K, Steward DJ. Developmental outcome in children undergoing surgery with profound hypothermia. Anesthesiology 1986;65:437–440.
37. Majnemer A, Limperopoulos, C. Developmental progress of children with congenital heart defects requiring open-heart surgery. Semin Pediatr Neurol 1999;6(1):12–19.
38. McCall RB. Environmental effects on intelligence: the forgotten realms of discontinuous non-shared within family factors. Child Development 1983;54:408–415.
39. Brandhagen DJ, Feldt RH, Williams DE. Long-term psychologic implications of congenital heart disease: a 25-year follow-up. Mayo Clinic Proc 1991;66:474–479.
40. Scarr S, McCartney K. How people make their own environments: a theory of genotypes-environment effects. Child Dev 1983;54:424–435.
41. Goldberg S, Morris P, Simmons RJ, et al. Chronic illness in infancy and parenting stress: a comparison of three groups of children. J Pediatr Psychol 1990; 15:347–358.
42. Casey FA, Sykes DH, Craig BG, Power R, Mulholland HC. Behavioural adjustment of children with surgically palliated complex congenital heart disease. J Pediatr 1996;21(3):335–352.
43. Glaser HH, Harrison GS, Lynn DB. Emotional implications of congenital heart disease in children. Pediatrics 1964;33:367–379.
44. Teeter PA, Semrud-Clikeman M. Child Neuropsychology: Assessment and Interventions for Neurodevelopmental Disorders. Needham Heights, MA: Allyn & Bacon, 1997.
45. Jonas RA, Newburger JW, Volpe JJ. Brain Injury and Pediatric Cardiac Surgery. Newton, MA: Butterworth-Heinemann, 1996.
46. Delamater AM, Brady EU, Blumberg MJ. Cardiac conditions. In: Brown RT, ed. Cognitive Aspects of Chronic Illness in Children. New York: Guilford Press, 1999;190–219.
47. Silbert AR, Wolff PH, Mayer B, et al. Cyanotic heart disease and psychological development. Pediatrics 1969;43:192–200.
48. Brunberg JA, Reilly EL. Doty DB. Central nervous system consequences in infants of cardiac surgery using deep hypothermia and circulatory arrest. Circulation (Suppl) 1974;49,50:II–60.
49. Majnemer A. Getting to the heart of the matter. Semin Pediatr Neurol 1999;6(1):1–2.
50. du Plessis AJ. Mechanisms of brain injury during infant cardiac surgery. Semin Pediatr Neurol 1999;6(1):32–47.
51. Baron IS, Fennell EB, Voeller KKS. Pediatric Neuropsychology in the Medical Setting. New York: Oxford University Press, 1995.
52. Glauser TA, Rorke LB, Weinberg PM. Clancy RR. Acquired neuropathologic lesions associated with hypoplastic left heart syndrome. Pediatrics 1990;85(6):991–1000.
53. Kirkham FJ. Recognition and prevention of neurological complications in pediatric cardiac surgery. Pediatr Cardiol 1998;19:331–345.
54. Mitchell D, Hilberman M. Brain function after hypothermic circulatory arrest. In: Hilberman M, ed. Brain Injury and Protection During Heart Surgery. Boston: Martinus Nijhoff, 1988;157–170.
55. Hesz N, Clark EB. Cognitive development in transposition of the great vessels. Arch Dis Child 1988;63:198–200.
56. Hickey PR. Neurologic sequels associated with deep hypothermic circulatory arrest. Ann Thorac Surg 1998;65:S65–709.
57. Miller G, Eggli KD, Contant C, Baylen BG. Myers JL. Postoperative neurologic complications after open-heart surgery on young infants. Arch Pediatr Adolesc Med 1995;149:764–768.
58. Morris RD, Krawiecki S, Wright JA, Walter W. Neuropsychological, academic and adaptive functioning in children who survive in-hospital cardiac arrest and resuscitation. J Learn Disabil 1993;26(1):46–51.
59. Mills S. Cerebral injury and cardiac operations. Ann Thorac Surg 1993;56:S86–91.
60. O'Dougherty M, Wright FS, Loewenson RB, Torres F. Cerebral dysfunction after chronic hypoxia in children. Neurology 1985;35:42–46.

61. Kaplan E. Introduction. In: Korkman M, Kirk U, Kemp S. NEPSY: A Developmental Neuropsychological Assessment. The Psychological Corporation. San Antonio: Harcourt Brace & Co., 1998;iii.
62. Korkman M, Kirk U, Kemp S. NEPSY: A Developmental Neuropsychological Assessment. The Psychological Corporation. San Antonio: Harcourt Brace & Co., 1998.
63. Wechsler D. Manual for the Wechsler Intelligence Scale for Children, 3rd ed. San Antonio: Psychological Corporation, 1991.
64. Spreen O, Risser AT, Edgell N. Developmental Neuropsychology. New York: Oxford University Press 1995.
65. Kaufman AS. Intelligence Testing with the WISC-III. New York: John Wiley, 1994.
66. Drotar D, Crawford P, Ganofsky MA. Prevention with chronically ill children. In: Roberts MC, Peterson L, eds. Prevention of Problems in Childhood: Psychological Research and Applications. New York: John Wiley, 1984.
67. Gardner E. Siblings of chronically ill children: towards an understanding process. Clin Child Psychol Psychiatry 1998;3(2):213–227.
68. Green M, Levitt E. Constriction of body image in children with congenital heart disease. Pediatrics 1962; 29:438–441.
69. Doucet SB. The young adult's perceptions of the effect of congenital heart disease on his lifestyle. Nurs Papers 1962;13:3–16.
70. Wollcott D, Norquist G, Busuttil R. Cognitive function and quality of life in adult liver transplant recipients. Transplant Proc 1989;12(3),3563.
71. American Academy of Pediatrics. Policy statement: screening infants and young children for developmental disabilities (RE9414). Pediatrics 1994;93(5):863–865.
72. Davis H. Counselling Parents of Children with Chronic Illness or Disability. Leicester: British Psychological Society, 1993.

Chapter 14

Chasing the Light: A Parent's Perspective on Hypoplastic Left Heart Syndrome

Louise E. Hall

In March 1996, a standard middle trimester scan revealed that my unborn baby had a malformed heart, later shown to be hypoplasia of the left ventricle. The following weeks, months, and years have seen us go through innumerable consultations, several cardiac catheters, and four operations. My son, Ben, is now 7 years old and is a happy, confident, normal boy who lives life to the fullest. As a result of my experience, in January 1997 I was asked to join the board of Little Hearts Matter, the group representing the interests of patients born with hypoplastic left heart syndrome in the United Kingdom. I initially joined the board as a parent contact for families that received a diagnosis of hypoplastic left heart syndrome antenatally, and was chair of the charity for over 2 years. I have thus had extended contact with very many families that have a wide range of differing experiences of living with a diagnosis of hypoplastic left heart syndrome. These include those such as myself, who have had a positive outcome subsequent to antenatal diagnosis, and those whose children have died, due to a termination of pregnancy, a choice of comfort care at birth, or before or subsequent to surgery. On our board, there are parents of children who have successfully undergone the various stages of the Norwood procedure, and one family whose son was successfully transplanted. We have recently extended our role to include all conditions with one functioning ventricle, and already have a trustee whose son has hypoplasia of the right ventricle.

This book discusses the surgical and cardiological management of children born with hypoplastic left heart syndrome. In terms of statistical probabilities or outcomes, 1 in 5000 pregnancies results in a baby with the syndrome. Of these, half are diagnosed antenatally. Half of the children born with hypoplasia of the left heart will be alive and well at the age of 5. Yet how does it feel to be one of those statistics? What does it mean to the parents and family that plunge from a world of eager expectation into a new, parallel world of medical jargon and surgical risk factors that most never knew existed? And how do parents deal with the fact that their child is alive only because of very recent advances in paediatric cardiology, and that those advances may not be enough for the future?

My son, Ben was diagnosed with hypoplastic left heart syndrome in 1996, after a routine 18- week antenatal anomaly scan indicated a likely cardiac problem. We were given the definitive diagnosis a few days later at a specialist centre, and then left to decide whether or not to continue with the pregnancy. Other parents often speak of their devastation, confusion, and anger at diagnosis:

> "We were in a state of complete shock and disbelief. My husband had clung onto the hope that "they'd got it all wrong" up to that point, so he found it terribly hard to cope with."
>
> Sarah, mother of Sophie

Being given the option of terminating my pregnancy was one of the most profoundly shocking moments of my life. We had some idea of the diagnosis before the scan, but had not anticipated that one of the possibilities was that there was no hope. Many families are touched by tragedy in lots of ways, but as medical expertise improves, the death, or, as in our case, the expected death, of a child is increasingly unusual. We do not, on the whole, have the tools as a society to deal with such absolutes. Babies aren't supposed to die. They may struggle for life through prematurity, or a congenital condition such as cystic fibrosis, but they don't die. We previously had faith in the medical community in a rather vague way, in that we believed it would be able to help when needed. When that help manifested itself as an offer to terminate the pregnancy, it was very difficult to come to terms with. An antenatal diagnosis of hypoplasia of the left heart offered us options that no parent should have to contemplate, as the following account eloquently illustrates:

> "The most difficult part about being told that our child had hypoplastic left heart syndrome was having to make decisions that we never dreamt in a million years we'd have to make. We had to deal with such HUGE questions. In the natural way of things this baby would die shortly after birth. Would it be right to put him through the pain and suffering of repeated surgeries, and ill health through all his life? Would it be more compassionate to end it now? I believe that modern medicine sometimes prolongs life when it shouldn't. Was this one of those cases? Was this a case when the availability of a safe and proper abortion was God's compassionate provision in an imperfect and painful world? Which would be the most painful—to have a termination of the pregnancy at about 23 weeks, or to give birth to our son and then possibly watch him die?"
>
> Joanne, mother of Benjamin

Out of all this turmoil, a decision has to be made and usually quite quickly. My husband and I did not hold any strong beliefs about abortion, but found that we could not face the possibility of terminating the pregnancy. We took this decision not primarily for the good of our baby, but for the sake of our own welfare, and that of the people closest to us. I am not sure even today if this was a selfish or selfless decision, only that we looked at what we thought we could cope with and chose that route. We did not know what we were condemning our child to face, but we did know that, just yet, we did not have the strength to confront his death.

The rest of the pregnancy was a never-ending round of scans and hospital visits. Neither of us had had any previous dealings with the medical world and, in many ways, those few short months before Ben was born were invaluable. We had the opportunity to learn about Ben's condition, to become used to the medical terminology, to get to know the team that would be looking after him, and to come to terms with the diagnosis and its ramifications. By the time Ben was born, we were prepared for the fact that he might die, but we were also prepared for the possibility that he might lead a full and normal life, which would be punctuated by various spells in hospital. This was mainly due to our own research, not as a result of listening to the medical professionals, who were almost universally downbeat. This is not unusual, even as increasing numbers of children survive the surgeries and lead happy, fulfilled lives, but the advent of the Internet means that many parents have access to previously unattainable knowledge about both the condition and its prognosis.

> "We saw photos on the Web site of happy, healthy-looking toddlers and children who had survived the surgeries. We saw that there was a possibility of a normal life ahead of us. We read how many of them behaved and played very normally—playing soccer, riding bicycles, going to school, and doing well. Yes, we did read of many tragic deaths, and a few strokes, and feeding difficulties, but we realised that this was not the whole picture."
>
> Joanne, mother of Benjamin

Much of this information can be misleading, of course, and it is true that there are not many Web sites devoted to how difficult it is bringing up a child with any cardiac condition, hypoplasia of the left heart notwithstanding, but it is sometimes vital to find something to hope for.

Hope, indeed, even in the face of such a devastating diagnosis, is the key. There were days during the latter stages of my pregnancy when I felt completely overwhelmed by the impact of what we had decided and what might happen because of that. The only thing that was left was the hope that, no matter what the outcome was to be, we would be able to deal with the aftermath. I was helped very much in this by my friendship with two parents of children with hypoplasia of the left heart, both of whose sons had died. Indeed, one of them died not long before Ben was born, so I was confronted in the starkest way possible with the realities of life with left heart hypoplasia. Nevertheless, their bravery, calmness, and, above all, normality gave me the hope that we, too, could overcome whatever was thrown at us.

> "There is no substitute for being able to share all your worries and concerns with somebody who has been down the same road before you."
>
> Joanne, mother of Benjamin

> "The most helpful thing we were offered was the chance to speak with other parents. It gave us hope."
>
> Sarah, mother of Sophie

I had the opportunity during my pregnancy to speak at length with one parent whose first child, Philip, had hypoplasia of the left heart, and who was awaiting the second stage of surgery. Her concerns, however, were like those of any new mother: How long should he sleep for? Was he feeding enough? Why was he crying? I was struck by how normal her worries were, even though the reasons behind them were so abnormal, and I realised that if Ben survived, we would essentially have, for us, a normal child with normal needs who would need a specific kind of care. Ultimately, even the strangest things can be absorbed into everyday life. Philip, for example, was fed, like many babies with hypoplasia of the left heart, by nasogastric tube. But this was not an issue, merely how he was.

We approached the birth of Ben with some trepidation, which was perhaps oddly tinged with excitement. We "knew" this baby better than many parents ever do and we wanted to meet him, even if it was only to say good-bye. He was born beautiful, pink, and of a good size at 3:20 p.m. on the 6th of August 1996, and underwent surgery 23 hours later. All those weeks and months of waiting culminated in 5 crucial hours in the operating room.

Many parents, of course, do not know about their child's condition until after he or she is born. One of the many small cruelties of hypoplasia of the left heart is that, when the baby is born, he or she looks as normal as any other baby. Any symptoms of cardiac failure can initially be attributed to other causes. Most babies thus present at paediatric cardiac centres in a poor condition, and decisions about surgery have to be made very quickly.

> "One factor contributed to the choice we made: our child was going to die anyway, so we had to give him the only chance available."
>
> Louise, mother of Alex

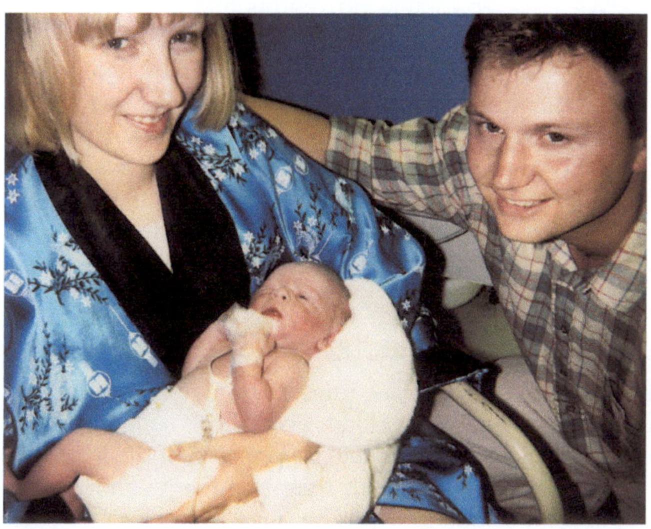

FIGURE 14.1. Ben just hours after his birth.

> "The doctor broke down as he began to tell us. Our little girl was too poorly to operate on."
>
> Judith, mother of Stephanie

Instead of being able to come to terms, to a greater or lesser extent, with the enormity of what they and their child are facing, these parents have to deal with the immediate reality of a diagnosis of hypoplasia of the left heart.

> "After the initial decision to go ahead, we wondered what we had done. What sort of quality of life would our son have?"
>
> Donna, mother of Oliver

The shock of seeing your newborn baby after he or she has had heart surgery is, like many other aspects of this condition, devastating. We handed over our "perfect" little boy at 23 hours old, and returned 5 hours later to a baby swamped by tubes, dressings, and endlessly bleeping machines.

> "When we got to see Alex, shock waves reverberated through my body. I felt my knees go weak and I felt physically sick. There were large strips of bandages that seemed to all but cover his face. He had too many tubes going into and out of his body to count. There seemed to be dried blood everywhere. He had dressings going down the entire length of his torso and what I now know to be residue iodine over any other exposed part of his chest. There were people hovering around the monitors constantly watching the screen and pressing buttons when alarms went off. It was only then that I realised the full horror of the fact that Alex's innocent little body had been cut open for this extensive surgery."
>
> Louise, mother of Alex

Living in the hospital can be quite a surreal experience. Whilst the sometimes frenetic, sometimes maddeningly boring, life in the wards goes on, "normal" life carries on outside. It occasionally breaks into the hospital routine in the form of visitors or domestic or work decisions that have to be made. Parents often have to support and help members of their extended family, but this can also be a positive experience.

> "My parents seemed to hurt all the more because I was their "baby" and they could feel my pain. They supported us both emotionally and practically whilst we were in hospital, which gave us time to learn to be a family."
>
> Louise, mother of Philip

After coping with surgery and its immediate aftermath, the next big hurdle is going home. Suddenly, the responsibility for the baby lies once more in the hands of the parents, especially so if the family has travelled some distance for surgery. As we were discharged, 2 weeks after Ben's surgery, his consultant told me, in all apparent seriousness, that we mustn't let him cry. He was 15 days old. We took the doctor at his word, however, and Ben hardly cried for the first 4 months of his life. We laugh about it now, but it highlights how potentially difficult taking such a sick baby home is. On the other hand, it puts an entirely different perspective on parenting.

> "We were aware that each and every moment with him was a blessing and a privilege. I never got frustrated or depressed by being woken frequently during the night, or being utterly exhausted."
>
> Joanne, mother of Benjamin

It is an immense strain knowing that the life of your child relies on a 2-mm Gore-Tex tube remaining patent. Most parents never contemplate the death of their child, but the parents of a child with hypoplasia of the left heart live with that possibility daily, especially in the early stages.

> "One of the hardest aspects was that at the back of my mind was either that Benjamin might die at any age—in the next day, the next week, a year or 10 years—or that he would have unpleasant and painful problems to face in the future, and that he might resent us for putting him through that."
>
> Bill, father of Benjamin

Gradually, a sort of normality is achieved. Sometimes, anticipated problems do not materialise, especially where siblings are concerned.

> "Both Jessica and Rachel are very loving and protective of Sophie. They have actually adapted and accepted much better than I thought they would."
>
> Sarah, mother of Sophie

Our eldest child, Sam, was almost 3 when Ben was born and is a sensitive, thoughtful, and caring child who often puts the needs of others before his own. Who knows if this is due to having a brother with such a serious problem, or whether he would have been like that anyway? Older siblings especially may have difficulty adjusting, but this is a recognised potential problem with the arrival of a new baby.

> "A couple of issues have arisen with Alex's older brother and sister, but nothing overly surprising."
>
> Louise, mother of Alex

The second stage of surgery does not engender the same fears as the initial operation, but is a stark reminder of how reliant we as parents are on the expertise of the surgeons. Most children go through this particular surgery with no problems, so it is much harder if difficulties do occur:

> "We thought we were over the worst, that he'd got through. We were starting to think about more than one day at a time, and then he died."
>
> Louise, mother of Philip

Death is an ever-present reality for the families of children with hypoplasia of the left heart. I went from knowing no one who had lost a child in infancy, to knowing several in a very short time. These parents face the reality of what we all fear:

> "It is hard to know that you will probably bury your child, and it was almost a relief when it actually happened, as if we were ready for it. We had a good $5\frac{1}{2}$ months with Philip and we have no regrets: we just know that life is not always the bed of roses we'd like it to be."
>
> Louise, mother of Philip

> "We felt so sad and angry that she had died, but also lucky that we had nothing to blame. We were left, however, with a paranoia about our other children."
>
> Judith, mother of Stephanie

> "We have absolutely no regrets about the decisions we made, and if we were given the chance to go back, we would do exactly the same thing again. The joy he brought us just in those few weeks far outweighs the pain and the suffering. We are so glad that we had the chance to hold him, to know him, to bring him home and have him live with us a little while. We wouldn't change that for anything."
>
> Jo and Bill Goldie, parents of Ben

As our children grow, we face the new challenges that all parents face: first day at playgroup, first nativity, first day at school. The third stage of surgery looms, ominously, like a dark cloud, and then is suddenly upon us. The luckiest children with hypoplasia of the left heart have minimal contact with their treatment centres after the second stage, so it can be quite a shock to be back

Figure 14.2. Sam, Freddy and Ben living a normal life.

in the midst of the medical world once more and, although recovery rate is good after surgery, the medium- to long-term prognosis for our children is unclear.

Having a child with hypoplasia of the left heart has a profound, sometimes unlooked for, effect on many aspects of family and, by extension, professional life. We have all faced the death of our children and this colours all kinds of decisions and choices that we make. The medical profession gives our children the chance to live, but it is we who must learn to live with the fears, uncertainties, and difficulties that are an inevitable result. It may be surprising, therefore, that some parents feel that it can be made into a positive experience:

> "When we look back at all the pain that we all went through, we now realise that our son was meant to be. He has given us a more positive outlook on life."
> Alison, mother of Jonathan

To see your child struggle for life is a humbling experience, but it can almost literally illuminate all that is good about being a parent. My husband and I had to think very carefully about why we wanted children and whether or not we could offer Ben a good quality of life: How many parents seriously do this in other circumstances? We now have a third, healthy son, Freddy, whose safe arrival was greeted with the mixture of joy and relief that only families with similar experiences could hope to understand.

It can be a dark road, with many pitfalls, but the light is worth chasing.

INDEX

MIX
Papier aus verantwortungsvollen Quellen
Paper from responsible sources
FSC® C105338

If you have any concerns about our products,
you can contact us on
ProductSafety@springernature.com

In case Publisher is established outside the EU,
the EU authorized representative is:
Springer Nature Customer Service Center GmbH
Europaplatz 3, 69115 Heidelberg, Germany

Printed by Libri Plureos GmbH
in Hamburg, Germany